ADVANCES IN AUTONOMOUS NAVIGATION THROUGH INTELLIGENT TECHNOLOGIES

AAP Advances in Artificial Intelligence and Robotics

ADVANCES IN AUTONOMOUS NAVIGATION THROUGH INTELLIGENT TECHNOLOGIES

Edited by

Niharika Singh, PhD

Thipendra P. Singh, PhD

Brian Azzopardi, PhD

AAP APPLE ACADEMIC PRESS

First edition published 2025

Apple Academic Press Inc.
1265 Goldenrod Circle, NE,
Palm Bay, FL 32905 USA

760 Laurentian Drive, Unit 19,
Burlington, ON L7N 0A4, CANADA

CRC Press
2385 NW Executive Center Drive,
Suite 320, Boca Raton FL 33431

4 Park Square, Milton Park,
Abingdon, Oxon, OX14 4RN UK

Library and Archives Canada Cataloguing in Publication

CIP data on file with Canada Library and Archives

Library of Congress Cataloging-in-Publication Data

CIP data on file with US Library of Congress

ISBN: 978-1-77491-700-8 (hbk)
ISBN: 978-1-77491-859-3 (pbk)
ISBN: 978-1-00353-305-4 (ebk)

AAP Advances in Artificial Intelligence and Robotics

The new book series *AAP Advances in Artificial Intelligence & Robotics* provides detailed coverage of innovations in artificial life, computational intelligence, evolutionary computing, machine learning, robotics, and applications. The list of topics covers all the application areas of artificial intelligence and robotics such as: computational neuroscience, social intelligence, ambient intelligence, artificial life, virtual worlds and society, cognitive science and systems, computational intelligence, human-centered and human-centric computing, intelligent decision making and support, intelligent network security.

With this innovative era of simulated and artificial intelligence, much research is required in order to advance the field and also to estimate the societal and ethical concerns of the existence robotics and scientific computing. The books in this series are practically relevant, so that the results will be useful for managers in leadership roles related to AI and robotics, researchers, data analysts, project managers, and others. Therefore, both theoretical and managerial implications of the research need to be considered.

The book series broadly considers the contributions from the following fields:

- Artificial Intelligence Applications in Security
- Artificial Intelligence in Bioinformatics
- Robot Structure Design and Control
- Artificial Intelligence in Biomedical and Healthcare
- Multi-Robot Intelligent Aggregation Mechanisms and Operation Platforms
- Artificial Intelligence and Learning Environments
- Advances in AI-Driven Smart System Designs
- Robot Navigation, Positioning, and Autonomous Control
- Robot Perception and Data Fusion
- Advances in Artificial Intelligence research
- Advances in AI-Driven Data Analytics and Innovation
- Application of Intelligent Systems for Solving Real-World Problems
- Hybrid Systems Design and Applications Using AI
- Robot Grabbing and Operation

- Robot Behavior Decision and Control
- Robot Motion and Path Planning
- Applications of Intelligent Systems and Computer Vision

For additional information, contact:

Book Series Editor: Subhendu Kumar Pani
Professor & Research Co-ordinator
Dept. of Computer Science and Engineering
Orissa Engineering College, Bhubaneswar, India
Email: pani.subhendu@gmail.com

BOOKS IN THE SERIES

Advancements in Artificial Intelligence, Blockchain Technology, and IoT in Higher Education: Mitigating the Impact of COVID-19
Editors: Subhendu Kumar Pani, PhD, Kamalakanta Muduli, PhD, Sujoy Kumar Jana, PhD, Srikanth Bathula, PhD, and Golam Sarwar Khan, PhD

The Fusion of Artificial Intelligence and Soft Computing Techniques for Cybersecurity
Editors: M. Ajabbar, PhD, Sanju Tiwari, PhD, Subhendu Kumar Pani, PhD, and Stephen Huang, PhD

Incorporating AI Technology in the Service Sector: Innovations in Creating Knowledge, Improving Efficiency, and Elevating Quality of Life
Editors: Maria Jose Sousa, PhD, Subhendu Pani, PhD, Francesca dal Mas, PhD, and Sergio Sousa, PhD,

Handbook of Research on Artificial Intelligence and Soft Computing Techniques in Personalized Healthcare Services
Editors: Uma N. Dulhare, PhD, A. V. Senthil Kumar, Amit Dutta, PhD, Seddik Bri, PhD, and Ibrahiem M. M. EI Emary, PhD

Fusion of Artificial Intelligence and Machine Learning for Advanced Image Processing, Data Analysis, and Cyber Security
Editors: Arun Kumar Rana, PhD, Rashmi Gupta, PhD, Sharad Sharma, PhD, Ahmad A. Elngar, PhD, and Sachin Dhawan, PhD

Artificial Intelligence and Machine Learning for Business
Editors: Namita Mishra, PhD, and A. V. Senthil Kumar, PhD, DSc

Advances in Autonomous Navigation through Intelligent Technologies
Editors: Niharika Singh, PhD, Thipendra P. Singh, PhD, and Brian Azzopardi, PhD

About the Editors

Niharika Singh, PhD
Lead Data Engineer, MPAC, Canada

Niharika Singh, PhD, has a decade of robust experience in technical consultancy, mentoring, and training. Currently, she is working as a Data Engineer in MPAC, Canada. Holding a PhD in Deep Learning (Computer Science Engineering) from UPES, India, she has significantly contributed to both academia and industry. She has an extensive publication record that includes research on email spam filtering, air quality prediction, path planning algorithms for networked robots, and more. She has also authored and contributed to books and book chapters on topics such as fog computing, autonomous navigation, etc. Her technical proficiency spans deep learning, data analytics, and cloud technologies, making her a distinguished figure at the intersection of academia and industry.

Thipendra P. Singh, PhD
Professor, School of Computer Science Engineering & Technology, Bennett University, Greater Noida, India

Thipendra P. Singh, PhD, is currently associated as a Professor of Computer Science and Dean of Academic Affairs at Bennett University, Greater Noida, India. Prior to this, he has been associated with UPES University and Sharda University, both in India. He has 27 years of rich experience. He is supervising PhD scholars with French and UK universities. He is a widely traveled academician and has participated in various platforms in several countries, including the UK, France, UAE, and Singapore. He has been the editor/author of 20 books on various allied topics of computer science and published more than 70 papers in various reputable journals. Dr. Singh is a senior member of IEEE and a member of various other professional bodies, including the Institution of Engineers (India), Association for Computing Machinery, EAI, International Society for Technology in Education, International Association of Engineers, etc., and is also on the editorial/reviewer panels of several journals. He has been a fellow of IETA-India since 2019. He is also on the boards of studies of several universities, both in India and abroad. He has been associated with many conferences throughout India as well as abroad as a chair, organizing committee member, technical program committee member, and session chair.

Brian Azzopardi, PhD
Founder and Chairman of The Foundation for Innovation and Research, Malta; Visiting Senior Lecturer and Consultant, University of Malta (UM), Malta

Inġ. Brian Azzopardi (BA), PhD, is the co-Founder and Chairman of The Foundation for Innovation and Research, Malta (FiR.mt), Senior Lecturer II at the Malta College of Arts, Science and Technology (MCAST), and Visiting Senior Lecturer at the University of Malta (UM). Since 2011, he has held senior academic and research positions in the United Kingdom, Lithuania, and Malta. He has also served in industry, government agencies, ministries, and research centers since 1998. He is the Coordinator of the Horizon 2020/Horizon Europe projects JUMP2Excel, NEEMO, TRANSIT, and PROMISE projects and has contributed directly to over €10M to affiliated institutions on projects with budgets over €30M over the past decade. He is a Senior Member of IEEE and a member of The Institution of Engineering and Technology, EI, Royal Society of Chemistry, and Chamber of Engineers. He is also the recipient of the MCST-JRC Young Scientist Award and STEM Young Researchers Award. In 2008, he received the Eur. Ing. title followed by the CEng and the EI Chartered Energy Engineer titles in 2012. In 2002, he received a BEng (Hons) from UM and a PhD from The University of Manchester in 2011. He also received teaching and pedagogical qualifications from MCAST (2008) and PGCHE from Oxford Brookes University (2012). Dr. Azzopardi is the editor and co-author of two books and over 100 research papers in peer-reviewed impact-listed journals and conferences, and is an invited speaker and the Chair of the MEDPOWER2022 Conference. He has supervised over 10 postgraduate master's, doctoral candidates, and postdoctoral students.

Contents

Contributors

Aryan
Department of ECE., J.C. Bose University of Science and Technology, YMCA, Faridabad, Haryana, India

Muhammad Yaseen Ayub
Department of Computer Science, COMSATS University Islamabad, Attock Campus, Pakistan

Katyayani Bhardwaj
Department of ECE, Indraprastha Engineering College, Ghaziabad, Uttar Pradesh, India

Arpita Jadhav Bhatt
Department of CSE&IT, Jaypee Institute of Information Technology, Noida, Uttar Pradesh, India

Maria Nawaz Chohan
National Defense University, Islamabad, Pakistan

Sana Nawaz Chohan
Foundation University Institute of Rehabilitation Sciences, Islamabad, Pakistan

Kanish Chugh
Department of ECE, Raj Kumar Goel Institute of Technology, Ghaziabad, Uttar Pradesh, India

S. B. Goyal
Faculty of Information Technology, City University, Menara City U, Malaysia

Shaurya Gupta
School of Computer Science, UPES, Dehradun, Uttarakhand, India

Rahul Gusain
Cybernetics Cluster, School of Computer Science and Engineering, University of Petroleum and Energy Studies, Dehradun, Uttarakhand, India

Ali Haider
Department of Computer Science, COMSATS University Islamabad, Attock Campus, Pakistan

Kingsley Igulu
Department of Computer Science, Ignatius Ajuru University of Education, Port Harcourt, Nigeria

Inam Ullah Khan
Department of Electronic Engineering, School of Engineering and Applied Sciences (SEAS), Isra University, Islamabad Campus, Pakistan; King's College London, UK

Sanjeev Kumar
School of Computer Science, University of Petroleum and Energy Studies, Dehradun, Uttarakhand, India

Ved Prakash Mishra
Amity University, Dubai, UAE

Wilson Nwankwo
Department of Computer Science, Edo State University, Uzairue, Nigeria

Friday Onuodu
Department of Computer Science, University of Port Harcourt, Port Harcourt, Nigeria

Mariya Ouaissa
Moulay Ismail University, Meknes, Morocco

Justice Palimote
Department of Computer Science, Rivers State Polytechnic, Bori, Nigeria

Ishika Punhani
School of Computer Science, University of Petroleum and Energy Studies, Uttarakhand, India

Jas Karman Saggar
School of Computer Science, University of Petroleum and Energy Studies, Uttarakhand, India

Devyansh Samyal
Cybernetics Cluster, School of Computer Science and Engineering, University of Petroleum and Energy Studies, Dehradun, Uttarakhand, India

Savita
Gurukul Kangri (Deemed to be University), Haridwar, Uttarakhand, India

Hitesh Kumar Sharma
Cybernetics Cluster, School of Computer Science and Engineering, University of Petroleum and Energy Studies, Dehradun, Uttarakhand, India

Hina Shoukat
Department of Computer Science, COMSATS University Islamabad, Attock Campus, Pakistan

Deepti Singh
Department of CSE&IT, Jaypee Institute of Information Technology, Noida, Uttar Pradesh, India

Niharika Singh
Lead Data Engineer, MPAC, Canada

Ravi Tomar
Senior Architect, Persistent Systems, India

Sona Tyagi
School of Computer Science, University of Petroleum and Energy Studies (UPES), Dehradun, Uttarakhand, India

Ankita Verma
Department of CSE&IT, Jaypee Institute of Information Technology, Noida, Uttar Pradesh, India

Sonali Vyas
School of Computer Science, UPES, Dehradun, Uttarakhand, India

R. K. Yadavd
Department of ECE, Indraprastha Engineering College, Ghaziabad, Uttar Pradesh, India

Noor E. Zahra
Department of Computer and Information Sciences, Princess Nourah Bint Abdulrahman University, KSA

Acknowledgments

Editing a book requires long and constant effort in pursuit of perfection. The editors are firstly thankful to all the contributors, reviewers, and academic friends who were involved in proofreading the texts.

The editors' next thanks goes to the publishing team for giving us the chance to publish this work and for the tireless efforts of their back-end team.

Editors Dr. Niharika Singh, Dr. Thipendra P. Singh, and Dr. Brian Azzopardi are thankful to their respective family members, colleagues, and all well-wishers who helped them in achieving this milestone. They also want to put on record their heartfelt appreciation for their organizations—the University of Petroleum and Energy Studies (UPES), the Bennett University, and the University of Malta—and their management to provide such a wonderful opportunity to let them grow academically.

The editors feel grateful to Prof. Ajith Abraham, *Founding Director, Machine Intelligence Research (MIR) Lab, USA,* Prof. Pramod Pathak, *Dean, School of Computing, University of Dublin, Ireland* and Prof. Kiran R., Dean, School of Engineering, Woxsen University, India for writing the foreword for this book.

Over and above, the editors feel thankful to the Almighty for showering His blessings upon them.

—***Editors***

Foreword

The rapid growth in the field of disruptive technologies has turned our day-to-day life inside out. Numerous research studies are being performed under the aegis of tech companies and research institutions, and it is becoming difficult to accumulate new information on a daily basis.

This book, *Advances in Autonomous Navigation through Intelligent Technologies,* edited by Dr. Niharika Singh, Dr. Thipendra P. Singh, and Dr. Brian Azzopardi**,** is an excellent collection of advanced work done in the field of autonomous navigation using technologies like AI, big data, Internet of Things, and communication networks. When the editors approached us to write a foreword for the book, we could not deny looking into the collection of material that ranges from establishing the fundamentals of intelligent autonomous navigation to applying these to the sophisticated systems in practice.

We are sure that this book will serve its intended purpose and will make a significant impact on the minds of readers. Also, some new thoughts will inspire future work in this field. We wish all contributors, authors and consuming fraternity a wonderful reading.

—Dr. Pramod Pathak
Dean, School of Computing, University of Dublin, Ireland

—Dr. Kiran R.
Dean, School of Engineering, Woxsen University, India

Preface

This book serves as a resource for academics, researchers, and anyone else interested in learning about and comprehending the many advancements in autonomous navigation. Computational intelligence and artificial intelligence should not be mistaken for each other, despite the fact that they are connected. The development, study, and application of numerous systems that can mimic human intelligence are fundamental to the science of autonomous navigation. Computing intelligence employs paradigms that are driven by biological or linguistic elements to solve computational problems. It is preferred over conventional and traditional mathematical modeling for tackling complicated real-world situations since the following criteria are met: (i) The problem involves a computing process that mathematical reasoning is unable to understand; (ii) The computational process is, by its very nature, decided at random (heuristic), and the computational problem contains some uncertainties. As a result, phenomena found in nature are copied, and the computer solutions they inspired are modified. Neural networks, fuzzy logic systems, evolutionary computation, social reasoning, artificial life, ambient intelligence, deep learning, and other similar concepts are now included in computational intelligence.

There are 13 chapters in this book, all of which present commonly encountered topics and issues and adapt various autonomous navigation solutions to them.

Chapter 1 lays down the foundation of book by introducing autonomous navigation. Autonomous navigation is a major issue in robotics and artificial intelligence. The process of assisting anyone from one location to another is called navigation. While responsive strategies specify the action to be performed at the specified time, systematic strategies aim to generate a complete action by specifying the action to be performed at a subsequent time step and using motion planning strategies. Systematic strategies aim to generate an entire action to the goal.

Chapter 2 focuses on the architectural side of autonomous navigation. It is necessary to have a directed and specialized software architecture to easily simulate its behavior and its linked components, such as perception, drive-by-wire system, object tracking, detection, and trajectory planning, with the modern advancements in autonomous navigation systems for self-driving automobiles and even robots. Vision, localization, environment modeling, logic and judgment, management, and other similar challenges must all be overcome to navigate independently.

Various algorithms for robot positioning and navigation are discussed in Chapter 3. A robot may require navigation to achieve its objectives, depending on its application area and goals. The military, transportation, industry, and space exploration are just a few of the many fields in which autonomous robots can be used. Mobile robots are capable of material handling, disaster relief, patrolling, and rescue operations. As a result, an autonomous robot that can freely move in both static and dynamic environments is required.

Methodologies for controlling the autonomous system is the subject of Chapter 4. An autonomous system (AS) is a network or group of networks that are all managed and controlled by a single body or organization. An AS is a diverse network that is typically managed by a significant business. Within an AS, several distinct subnetworks share common routing rules and algorithms. The detailed description around this makes the subject of this chapter.

Chapter 5 presents a comprehensive study of comfort and motion planning for highway autonomous driving. Autonomous vehicles are becoming increasingly common across numerous nations. In terms of motorized manufacturing and the research that goes along with it, the locomotive industry represents one of the most promising developments. The Houdina-Radio Control Company made the first contributions to driverless vehicles. Later, it effectively continued with its demonstration using controlled radio signals and a sprawling motor vehicle. Self-driving car testing in urban traffic management has been legalized in several developed nations, including the United States. The application and use of ADAS have been accepted into computerized driving systems by the UN Automobile Conventions Act.

A detailed review on self-learning routing protocols for autonomous unmanned aerial networks (A-UAN)" is presented in Chapter 6. UAVs have already been used in a variety of civil and military settings. UAVs are used to minimize human intervention, particularly in dangerous areas where human life is at risk. For instance, UAVs can be used to deliver medical aid in remote areas, monitor volcanic eruptions, or monitor other areas where human physical access is impossible, such as in a combat zone and many other types of monitoring. For drones to continue providing uninterrupted service, it is essential to develop a secure and rapid communication mechanism between them given the increased use of UAVs.

Chapter 7 includes a survey on smart intrusion detection-enabled flying Ad hoc networks (FANETs) in healthcare. The chapter proposes the Internet of Medical Things (IoMT), combining IoT and medical technology. It highlights the use of flying ad hoc networks (FANET) in healthcare for communication and transportation via unmanned aerial vehicles (UAVs). It emphasizes IoT's role in patient record accessibility and the need for robust cybersecurity measures.

Chapter 8 discusses the concepts and algorithms of mobile devices. The market value of mobile devices has significantly increased as a result of their widespread adoption for personal and professional use. Nowadays, a wide range of mobile devices is used by people, each with its design, set of features, and constant improvement. Participating in mobile technology is using a mobile phone to make and receive phone calls, send text messages, and access the Internet. It is the channel through which data travels from one system to another. Wireless two-way communication systems, portable computers, and networking hardware are all examples of mobile technology.

The focus of Chapter 9 is on sensors, IoT, and other related components required for autonomous navigation.

Various issues encountered in autonomous navigation are explained in Chapter 10. The rapid development of technology in today's world has made it possible to make significant and rapid changes to our day-to-day lives. It's because more people are using devices with internet access, which generates a lot of data, or because businesses are competing to offer their products and services to the market in the best possible way. One of these technologies' byproducts, autonomous navigation systems (ANS), has piqued the interest of the automotive and robotics industries. The primary feature of ANS is its ability to navigate these vehicles with minimal human intervention. A self-driving car manufactured by the well-known automaker Tesla in the United States is one of the most well-known examples of ANS.

Chapter 11 discusses the current state of art of autonomous navigation and throws some light on its future. When it comes to assisting humans in a variety of real-world situations, autonomous systems can be of critical importance. This could be used in many different ways, including humanoid robots, driverless cars, military systems, grocery store carts, delivery robots, agricultural machines, and so on. To select a single path and navigate its surroundings, the autonomous navigation system requires no human intervention at all or only a small amount of it. These systems are currently also used to help people in difficult situations, such as those where they cannot go.

Most of the recent research focuses on autonomous mobile robots navigating unknown environments. To achieve this, effective path planning is vital for obstacle avoidance and successful target achievement. Robots are categorized into manual, remote-controlled, and autonomous systems. Autonomous robots exhibit human-like decision-making capabilities, relying on perception, analysis, and action within their surroundings. Chapters 12 and 13 discuss the environmental learning frameworks and analysis of various algorithms used for the purpose of automatic navigation.

The aforementioned chapters are, in fact, cutting-edge pieces of research in the field of autonomous navigation. Overall, the book presents an interesting discourse on the topic in its entirety, and we expect that it will be useful to the practitioners of this field.

—*Editors*

CHAPTER 1

Introduction to Autonomous Navigation

NIHARIKA SINGH,[1] ISHIKA PUNHANI,[2] NOOR E. ZAHRA,[3] and RAVI TOMAR[4]

[1]*Lead Data Engineer, MPAC, Canada*

[2]*University of Petroleum and Energy Studies, India*

[3]*Department of Computer and Information Sciences, Princess Nourah Bint Abdulrahman University, KSA*

[4]*Senior Architect, Persistent Systems, India*

ABSTRACT

The path-planning problem for robots has been an active analysis area for several years that started in the mid-1970s. There were different strategies proposed for this research problem. It is essential to select a suitable method to achieve both the quality and efficiency of a search. Furthermore, it is important to avoid all identified obstacles in the area. The mobile robot uses a heuristic approach to interact with obstacles. The main goal of mobile robots is to reach predefined target locations without encountering barriers. Robotic applications require an intelligent selection of the optimal path through exploration and learning of the environment. Simultaneously, robots should reach the goal points at specific discrete times by avoiding obstacles. The study introduces an efficient navigation system for robots. The main contribution of the study was exploring the environment for robot navigation. The objective of this system becomes complex due to the optimal sets of actions during automation. The different interactive components necessitate the parallel process of learning the environment.

Advances in Autonomous Navigation Through Intelligent Technologies.
Niharika Singh, Thipendra P. Singh, Brian Azzopardi (Eds.)

Ultimately, the system's output aims to provide collision-free navigation in an unknown environment.

1.1 INTRODUCTION

Autonomous navigation is a critical problem in robotics and artificial intelligence. Navigation is the process of assisting movement from origin to destination. Real-time navigation is a simple activity for individuals or other animals, but it is exceptionally hard for robots, especially in an unknown and uncertain environment [1]. Environment learning is a crucial component in autonomous navigation. There are two kinds of robot navigation based on their surroundings, as described by Zuo et al. [2]:

1. **Navigation in a Static Environment:** The robot knows the global route and the position of the obstructions ahead of time.
2. **Navigation in an Unfamiliar Environment:** The robot has no knowledge of the path or the location of the hurdles. This classification is further subdivided as follows:
   i. **Approach based on Behavior:** The environment can be interpreted, and the trail can be obtained within the sensor threshold.
   ii. **Learning-based Approach:** The robot learns the navigation approach on its own, and performance is improved with training. Implementing self-learning capability (deep reinforcement) within an autonomous system [9] necessitates a learning-based approach. The ability of a system to self-learn allows for rational decision-making. The system recognizes its surroundings and makes decisions about its navigation.

Thus, autonomous navigation can be described as the development and execution of a plan from one space location to another in order to complete the task effectively while detecting and avoiding clashes with obstacles or inappropriate actions. To optimize both energy and cost metrics, the plan should make the best use of the available resources. The navigation module for self-driving vehicles is critical to consider. Robots are used in nearly every single industry where a recurring and challenging task is involved, as well as tasks that are either too dangerous or impossible to do manually, such as: (i) painting the car, welding various pieces or machine components, and surface finishing in the aerospace and automotive industries. (ii) Space and submarine tools. (iii) Catastrophic fritter away restoration in administration laboratories, nuclear services, and medical laboratories. (iv) Inspection of parts. (v) Assembling consumer goods and electronics. (vi) Examination and allocation of components in numerous industries. In industrial engineering and advanced technology, the concept of the supremacy of mobile robots includes numerous technological fields. These methodologies are aimed at trajectory planning, collision avoidance, mobile robot localization, route planning [3], and many more. Almost entirely, the perception of an autonomous mobile robot's layout planning, obstacle avoidance [4], and navigation job relies on the ease of using an explicit demonstration of the navigation environment. Within the fields of robotics, the following paradigms [5] have been revisited for autonomous navigation:

- hierarchical (plan-based approach);
- reactive;
- hybrid deliberative.

The aforementioned paradigms point to a substantial paradox in Autonomous Navigation. The plan-based approach utilizes a global overview, based on global self-localization, and is distinguished by the sense-plan-act cycle. The reactive method incorporates local control laws that relate to feature points, dependent on precise local feature identification, and is distinguished by sense-act. If the hierarchical approach employs a sense-plan-act strategy and the reactive approach employs a sense-act strategy, then the hybrid-deliberative process is defined by Plan, Sense-Act. As a result, this method integrates the finest of both worlds in that it includes a planner and the deliberate part of the planner. Following an overview of the paradigms, the following section examines the primary operational characteristics of an autonomous vehicle. The following are the primary operational features of an autonomous vehicle: (i) observation; (ii); intellect; and (iii) action.

Artificial intelligence is a perfect ingredient because it embodies the aforementioned traits. Numerous tools and algorithms are available in the field of artificial intelligence. Hill Climbing, A*, D*, Fuzzy logic [8], Artificial Neural Network (ANN) [11], Neuro-Fuzzy, Genetic Algorithm, Swarm Intelligence, particle swarm optimization (PSO), and a long list of other AI algorithms are just a few examples of the many algorithms used in AI. However, reprogramming is necessary in the aforementioned ways if the environment even slightly changes. Any autonomous mobile system must self-learn (use reinforcement learning) to prevent it. As a result, the study demonstrates that artificial intelligence is increasingly serving as an autonomous vehicle's brain. This chapter focuses on introducing artificial intelligence (AI) before classifying AI and its applications.

1.2 BACKGROUND

John McCarthy coined the term "artificial intelligence" to describe a machine's ability to perform functions that, if performed by a human, would be considered intelligent, such as reasoning, learning, decision-making, adaptation, control, and perception. Because of its pyrotechnic and commercial appeal, this definition of machine intelligence is hotly debated. According to the research, robots can be classified into three types: manually controlled, remote-controlled, and autonomously controlled robotic systems, as illustrated in Figure 1.1. The autonomous robot should be able to navigate in an unknown environment without the assistance of humans. The robots must have proper path planning [12] to avoid colliding with obstacles and reach the target. In the recent past, there have been numerous applications for the design, development, and operation of autonomous navigation in an unknown environment. A manual robot is an automated system that requires extensive human involvement in order to function. These automated systems require exceptional human authority to control the various processes that are integrated with the machines. Manual robots encompass a range of robotic systems, from the most basic to the most advanced, each dealing with a specific control system based on its diligence. A remote-controlled robot, on the other hand, is defined as any vehicle that is teleoperated by a means that does not restrict its motion with an origin external to the device. Furthermore, the last and most important type of robot is autonomous. Autonomous robots are human-like robots that have the ability to reach conclusions and carry out actions. The primary working mechanism of these robotic systems is to perceive the environment, draw conclusions based on that perception, and then initiate movement within that environment. In terms of mobility, for example, these decision-based actions include, but are not limited to, the following fundamentals: beginning, ending, and maneuvering around obstacles in their path. Robots are further classified as programmable, non-programmable, adaptive, and intelligent. A programmable robot is a unique robot with actuators on each joint. Depending on the type of use permitted, the robots may be reprogrammable. After the robot has been modified once to play out power in the given example and fixed arrangement, the capacity and utilization of the robots can be changed by reinventing. The primary disadvantage of this autonomous robot is that it cannot be customized. It continues to operate regardless of whether its assignment needs to be changed (in case of an emergency). A non-programmable robot is one of the most basic types of robots. This robot is not a robot, but rather an exploiter lacking a reprogrammable control device. Robots are frequently added to programmable devices used in large-scale manufacturing in businesses. Adaptive robots are industrial robots that can adapt to different ranges in the process independently. These robots, however, are more sophisticated than programmable robots. They can be adapted to some extent, and after evaluation, they can perform the required action in the adapted area. Intelligent robots, as the name implies, are the most brilliant of all types of robots, equipped with sensors and microprocessors for data storage and processing. Because of their situation-based analysis and task-performance abilities, these robots perform extremely well. Intelligent robots can detect pain, smell, taste, vision, and hearing. They can perform actions and expressions such as emotions, thinking, and learning accordingly.

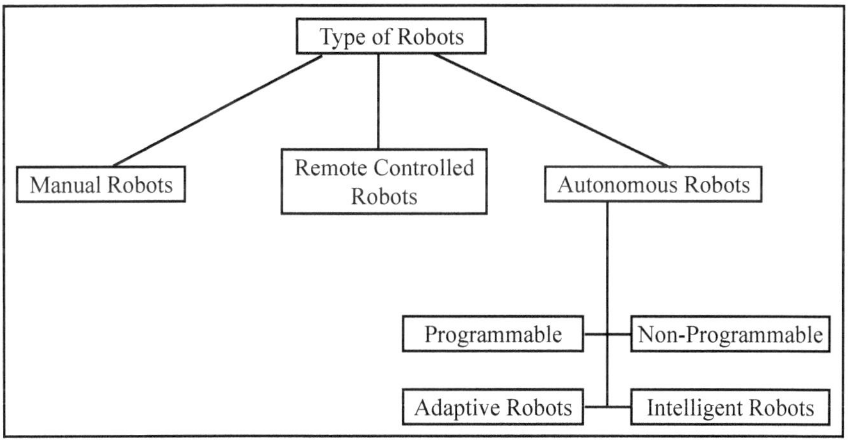

FIGURE 1.1 Types of robotic systems.

1.3 INTRODUCTION TO ARTIFICIAL INTELLIGENCE

Artificial intelligence (AI) is an upcoming paradigm shift in the field of computer science, providing computers with the discretion and ability to perform tasks that typically require manual or human intervention. AI can handle highly complex tasks efficiently and deliver results within a specific timeframe. With the development of machine learning and deep learning within the realm of artificial intelligence, machines can now execute actions that previously demanded years of experience and undivided attention, with just a few lines of code providing instructions based on past data. AI heavily relies on complex algorithms, which are a series of specific commands that guide the machine to produce a desired output. AI algorithms primarily depend on a continuous flow of data to help the model learn and adapt to patterns within the data. These algorithms are capable of recognizing patterns that may be too intricate for the human brain to comprehend. At times, the data is so vast that it needs to be represented in tens or hundreds of dimensions in vectors. This method is not easily understandable for the human brain, as it cannot process such large amounts of data simultaneously. AI is heavily reliant on data, making it essential for every industry. Data, once considered useless up until 70 years ago, now plays a significant role in helping companies and enterprises develop new business opportunities. These companies can quickly determine how, when, and where customers are likely to use a service, making it conveniently available, which has contributed to the success of companies like Uber. Until the past decade, only major corporations, which had the resources, could use Artificial Intelligence. However, with improved public technology, artificial intelligence was democratized for the public. Now everyone has the power of artificial intelligence built into their smart mobile devices. Applications like photo editing and file management are using artificial intelligence to provide ordinary people with a better experience and increased productivity right at their fingertips. With increasingly powerful machines at the disposal of the typical person and rapid research in artificial intelligence, the lives of ordinary people will be nothing but easier and more convenient.

1.3.1 CLASSIFICATION OF ARTIFICIAL INTELLIGENCE

Unlike many other branches of computer science, artificial intelligence spans almost all possible applications through myriad data and structures. Broadly, AI has two types, but it can be further classified into more subdomains based on its functionalities. One of the broad classifications is weak artificial intelligence or narrow artificial intelligence. This type of AI focuses on a single task and directs all its efforts towards achieving one goal: acquiring the best performance on that task. The other type of AI is Strong AI, an artificially intelligent machine that can think and process information like humans. Every weak AI is helping to achieve the goal of building a Strong AI model someday. Another type of classification for AI is based on its functionality. For example, a machine can be a reactive machine where the device does not have the capabilities to remember past data and makes decisions as the data comes through. These machines were good in the past, but with the development of better hardware, we now have limited memory machines that can store some data to make valuable decisions in the not-so-distant future. Autonomous vehicles use this learning in many areas like observing lane changes and so on. Then there are two more proposed models: theory of mind, which can understand emotions, beliefs, thought processes, etc., and the last type of machines are the self-aware ones. These machines have their consciousness and their own set of beliefs and perceptions.

1.3.2 APPLICATIONS AND CONCERNS OF ARTIFICIAL INTELLIGENCE

Artificial Intelligence has been enormously pervasive in most industries, products, and services that surround us today. Applications range from producing information-altering E-Systems like deep fakes, which work on Generative Adversarial Networks (GANs), to autonomous vehicles and places like creating original music and movies. While AI applications have since proved to be a massive boon for the entire community, it has many challenges and hurdles. The first and foremost is the problem of ethics. AI models are capable of learning about anything from the data they are being fed. Since these models do not have discretionary power, they cannot decide what is ethically correct and wrong. Applications like deep fakes and GPT-3 have since been brought to light due to their uncanny abilities to alter accurate data into some truth-altering information, which is seemingly indistinguishable for the human eye. Although they have the power to solve many problems, these applications seem to have started creating new ones. Ethics is not the only challenge we face with AI. Cognitive abilities, knowledge representation, resource planning, social intelligence, perception, general intelligence are just some of the other difficulties AI faces. However, with increasingly successful research outputs and the continuing efforts towards making AI possible for everyone, AI is becoming better and better and is coming closer to becoming one single pool of information for all the needs.

1.3.3 APPLICATIONS OF AI IN AUTONOMOUS NAVIGATION

Autonomous vehicles rely heavily on substantial data dumps by the onboard sensors to make decisions in real-time. However, this data requires almost instant processing. The algorithms must make insights and understand the data as humans do. This cognitive touch to these algorithms is provided by artificial intelligence. Almost every AV manufacturer and researcher has to build models using branches of AI like machine learning and deep learning to make sense of this data and provide impromptu results and feedback to the system for path planning operations. While complete autonomy over human interactions on vehicles is a far-fetched dream, AI proves helpful in driving assistance in emergency cases, taking over as a co-pilot in the car in urban planned settings, traffic light coordination, cross traffic relays, and simulating blind spot detection. With a massive amount of pictorial data, the AI models have started learning to adapt to complex environments with exponential numbers of stakeholders and objects in consideration. A large area where AI is also helping autonomous vehicles is keeping the vehicle's health in check. With real-time data on the vehicle's physical conditions and breakdowns, AI is reaching a new application area.

KEYWORDS

- **algorithms**
- **artificial intelligence**
- **autonomous navigation**
- **environment learning**
- **E-systems**
- **unknown environment**

REFERENCES

1. Zhu, A., & Yang, S. X., (2007). Neurofuzzy-based approach to mobile robot navigation in unknown environments. In: *IEEE Transactions on Systems, Man, and Cybernetics, Part C (Applications and Reviews)* (Vol. 37, No. 4, pp. 610–621).
2. Zuo, B., Chen, J., Wang, L., & Wang, Y., (2014). A reinforcement learning based robotic navigation system. *Conference Proceedings IEEE International Conference on Systems, Man and Cybernetics, 2014*, 3452–3457.
3. Alempijevic, A., (2004). High-speed feature extraction in sensor coordinates for laser rangefinders. *Australian Conference on Robotics.*
4. Borenstein, J., & Koren, Y., (1989). Real-time obstacle avoidance for fast mobile robots. *IEEE Transactions on Systems, Man and Cybernetics, 19*(5), 1179–1187.
5. Charles, R., John, W., et al., (2014). High-speed autonomous navigation of unknown environments using learned probabilities of collision. *IEEE Conference on Robotics and Automation.*
6. Daniel, S., Patrick, S., et al., (2010). Vision-based autonomous navigation system using ANN and FSM control. *Latin American Robotics Symposium and Intelligent Robotics Meeting.*
7. Jefferson, R. S., Gustavo, P., et al., (2011). Vision-based autonomous navigation using neural networks and templates in urban environment. *1st Brazilian Conference on Critical Embedded Systems.*
8. Meng, W., & James, N. K. L., (2008). Fuzzy logic based real-time robot navigation in an unknown environment with dead ends. *Robotics and Automation Systems.* Elsevier.
9. Benjamin, L., Raja, C., et al., (2004). On fault tolerance and robustness in autonomous systems. *Proceedings of the 3rd IARP-IEEE/RAS EURON Joint Workshop on Technical Challenges for Dependable Robots in a Human Environment.*
10. Benjamin, L., Alexander, L., et al., (2004). Fault tolerance in autonomous systems: How and how much? *Proceedings of the 4th IARP-IEEE/RAS EURON Joint Workshop Technical Challenges for Dependable Robots in a Human Environment.*
11. Katal, A., & Singh, N., (2022). Artificial neural network: models, applications, and challenges. In: Tomar, R., Hina, M. D., Zitouni, R., & Ramdane-Cherif, A., (eds.), *Innovative Trends in Computational Intelligence.* EAI/Springer Innovations in Communication and Computing. Springer, Cham. https://doi.org/10.1007/978-3-030-78284-9_11.
12. Singh, N., Prateek, M., & Chauhan, P., (2019). An efficient path planning algorithm for networked robots using modified optimization algorithm. In: *International Journal of Innovative Technology and Exploring Engineering* (Vol. 8, No. 12, pp. 2738–2743). Blue Eyes Intelligence Engineering and Sciences Engineering and Sciences Publication – BEIESP.

CHAPTER 2

Architecture of Autonomous Navigation

NIHARIKA SINGH,[1] SONA TYAGI,[2] and VED PRAKASH MISHRA[2]

[1]*Lead Data Engineer, MPAC, Canada*

[2]*School of Computer Science, University of Petroleum and Energy Studies (UPES), Dehradun, Uttarakhand, India*

[2]*Amity University, Dubai, UAE*

ABSTRACT

With the modern advances in autonomous navigation systems for self-driving automobiles and even robots, it is necessary to have a directed and specialized software architecture in order to simulate their behavior and its linked components, such as perception, drive-by-wire system, object tracking, and its detection, and trajectory planning, with ease. In order to navigate autonomously, one must solve a variety of difficult challenges in areas as diverse as vision, localization, modeling of the environment, logic and judgment, management, etc. It is crucial that these features are created and combined into one conceptual architecture. The primary characteristic of the routing strategy is its capacity to determine optimal mobility choices in actuality, thereby accounting for the severe restrictions imposed by the types of settings studied. Due to the limited knowledge and high dynamic of such settings with moving objects, motion safety is a crucial concern (other automobiles, pedestrians...) [1]. Machines equipped with artificial intelligence (AI) are designed to emulate human behavior.

Advances in Autonomous Navigation Through Intelligent Technologies.
Niharika Singh, Thipendra P. Singh, Brian Azzopardi (Eds.)

2.1 INTRODUCTION

This guidance system can be broadly divided into systematic (also known as mobility scheduling) and responsive strategies from the perspective of motion determination. While responsive strategies specify the motion to be done at the specified time interval, systematic strategies strive to generate a full motion specify the motion to be done at a subsequent time step, systematic strategies strive to generate an entire movement together all the way to the objective by utilizing motion planning strategies. A motion planning problem must be solved using deliberate methods. Because of their inherent complexity and the fact that the automobile only has a certain amount of duration to decide on its next direction to travel, they require an environment model that is as accurate as feasible. Reactive methods, on the other hand, can function online utilizing local sensor data; they may be applied in any environment, regardless of how unfamiliar, changing, or dynamic it may be [2].

Every day that goes by, the world makes impressive technological and automated progress. Smart cities (SG) and intelligent vehicle communication networks (IVCN) are made possible by integrating intelligent industrial area networks (IIAN), intelligent household area networks (IHAN), and smart grids. The Internet of Vehicles (IoV), Vehicle to Everything (V2X), Vehicle to Vehicle (V2V), and Vehicle to Infrastructure's intelligent node for autonomous vehicles includes the essential enabler of IVCN (V2I). Since 1920, when research on autonomous driving first began, several developments have been made. But even with a certain amount of intelligence, technology still needs human assistance [3].

Although several businesses, like Tesla, Google, and Uber, have made significant investments in the development of this technology, the autonomous system is still a very active research area because of its enormous hurdles. An excellent autonomous system is able to make wise judgments in actual situations. Improved localization, perception, and detection algorithms remain top research priorities [2].

The following are the key issues on which autonomous vehicle technology is based:

- Where am I right now?
- Where am I surrounded with?
- What will take place next?
- What has to be done?

The localization issue arises from the initial query, "Where am I right now?" In the existing environment, the ability of the automobile to detect or locate itself. The following question concerns perception and is about learning about the environment. What will take place next? The third inquiry, which is based on the information experienced or detected, is the domain of the forecast about the environment. "What has to be done, should be done, what should be done, what should be done?" These cars are dependable and safe to drive because they utilize a variety of sensors and algorithms to solve all these core problems [2].

Multiple sensors positioned on the automobile components are used by autonomous cars to perceive their environment. Following the receipt of data from these sensors, decisions are made on the safest route to get to the desired location, taking into account the best travel times and distances. More cutting-edge approaches, including location, item recognition and classification, route planning, and knowledge discovery obtained from several sensors, are required to complete the task (Figure 2.1).

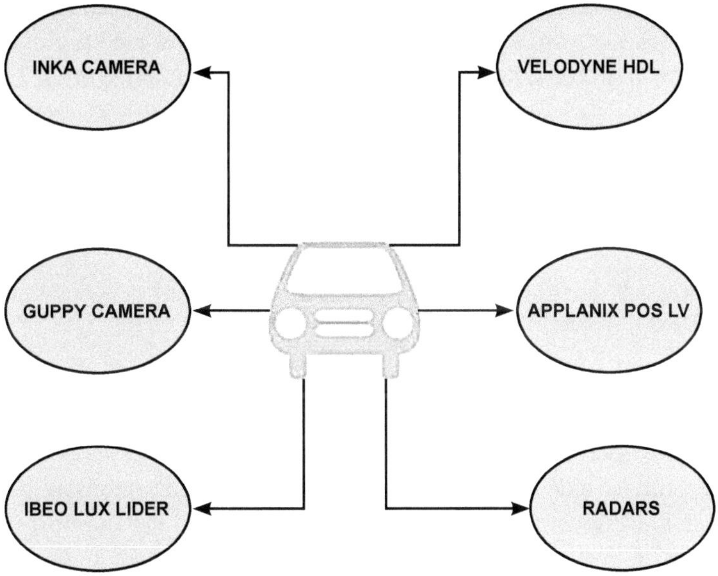

FIGURE 2.1 Sensors in automobiles.

In order to overcome these challenges and get the best performance, a branch of artificial intelligence called deep learning is becoming extremely popular due to the availability of highly capable computing tools like graphics processing units (GPUs) and a large quantity of data.

The performance of AVs has been enhanced by deep learning algorithms since they guarantee precision and quick processing [4].

2.2 NAVIGATION ARCHITECTURE

The autonomous navigation architecture may be further broken down into four stages: modeling or decision-making, localization, path planning, and lastly, path tracking. A mobile vehicle path is defined as the path the machine would take in a given environment. It is crucial for the execution of mobile robot missions. Path-tracking controllers are utilized to carry out path-following tasks, and an effective path-tracking controller achieves a minimum lateral distance as well as the direction that the vehicle takes to follow the prescribed route. Their major objective is to steer and guide the robot independently along a path while continuously producing speed and steering orders to correct tracking mistakes, which are mostly caused by variations in both direction and distance from the route of the automobile. For this, feedforward and feedback control techniques are employed, with a trade-off between control error and control effort [2] (Figure 2.2).

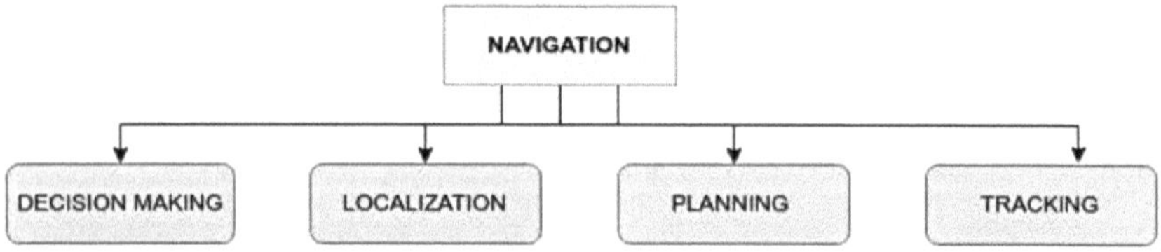

FIGURE 2.2 Various processes of navigation.

2.2.1 DECISION-MAKING ARCHITECTURE

In order to process the observation data obtained from the sensors installed on the vehicle, independent judgment is necessary for automobiles. These insights aid the computer in the automobile in making the best choices. Both end-to-end learning approaches and the combined perceive, plan, and act method may be used to calculate these decisions. The end-to-end technique maps sensor data directly, without the need for any intermediary processes, to control outputs. In Figure 2.3, an Artificial Intelligence-based automobile is displayed. Each stage in the perceive-plan-act approach used by AVs can be carried out using either traditional methods that need no learning or the most recent Deep Learning and Artificial Intelligence techniques, as shown in Figure 2.3. The end-to-end execution strategy always makes use of Deep Learning methods. Different combinations of learning and non-learning approaches are possible; for instance, a deep learning-based object detector feeds information into the A algorithm, which is used for designing the route [2, 3] (Figure 2.4).

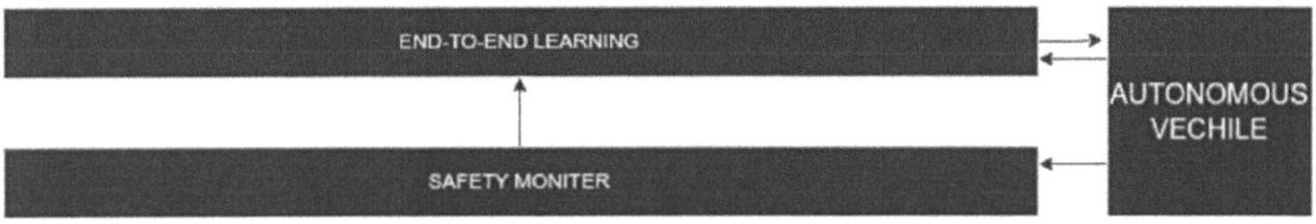

FIGURE 2.3 Navigation architecture using end-to-end learning.

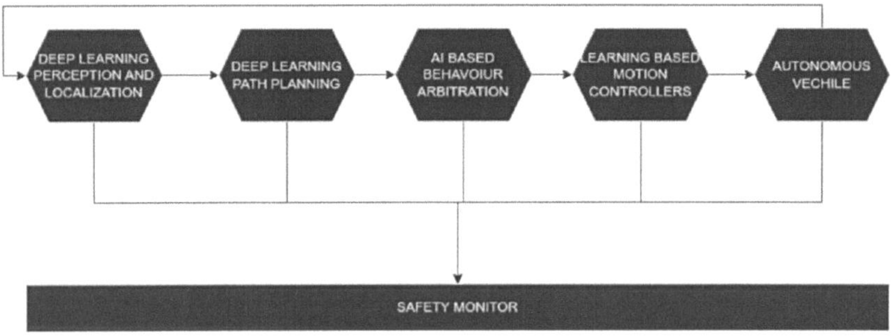

FIGURE 2.4 Deep learning-based safety monitoring in AVs.

2.2.2 PERCEPTION AND LOCALIZATION

3D perception is often chosen for a better comprehension of the environment. Only a 2D world can be captured in images produced by cameras. In general, 3D perception uses LiDAR sensors. The field of vision, resolution, range, and rotation/frame rate of the LiDAR system are used to gauge its performance. Another 3D sensor with a 360° field of view is Velodyne. Autonomous cars cannot afford any delays in information transfer; hence, a range of at least 200 m is necessary to guarantee processing at very fast rates.

The following are the environments that autonomous cars can operate in: Multiagent environments, dynamic, unknown, stochastic, sequential, and partially observable are just a few examples. Due to all of these environmental factors, autonomous driving is a very difficult task. Cars should be able to recognize any circumstance, including people, drivable places, and other agents in the environment. Operating in urban surroundings, where various items occur and blockings are quite frequent, makes the work harder and more difficult [2].

An important aspect of observing the environment is the use of deep neural networks. Various deep neural network techniques for object recognition have been developed, where objects are considered as 2-dimensional regions of attention. DNNs are employed in several additional research areas to perceive the environment using pixel-by-pixel picture segmentation, 3D bounding boxes from LiDAR, and, occasionally, 3D representations of objects using LiDAR camera combined data. On a jovial side, picture data might be helpful for item identification. However, complex information about the scene is lost when calculating the 3D locations of objects from 2D photos. The following are the two most often used techniques for identifying traffic scenes:

- Instance segmentation; and
- Object detectors.

Semantic and instance segmentation are crucial for secure navigation and understanding the environment. For misclassifications at the pixel scale in photographs, a failure identification system called FSNeT is recommended. The framework for effective knowledge distillation using transformers is provided for the semantic segmentation of driving scenarios on roads. For instance, segmentation, a convolutional neural network approach utilizing multiscale attention is presented [5] (Figure 2.5).

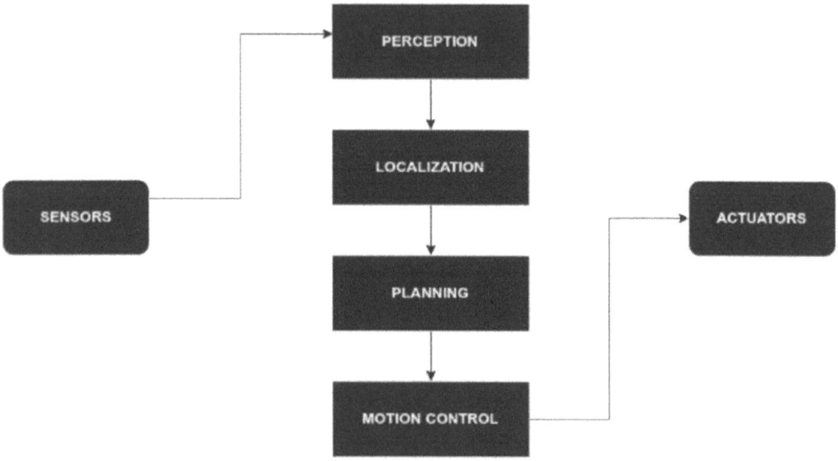

FIGURE 2.5 Flow of control from sensors to actuators in AVs.

2.2.3 *LOCALIZATION*

An essential component of navigation is localization. The localization challenge is determining the vehicle's pose (alignment + location) as it moves through the environment. Some of the most current AV research developments include Deep Learning-based techniques, which are interesting to notice that make end-to-end driving choices based solely on sensor data, eliminating the requirement for localization and mapping. The behavior reflex method is what is used in this. The most popular method for localization in autonomous cars is GPS. In order to account for signal loss in the event of any potential outage, different sensor data is merged with GPS data. There are several sensor fusion methods described in the literature. The standard methods for sensor fusion that are most frequently employed include the Kalman filter, unscented Kalman filter, extended Kalman filter, multimodal Kalman filter, and particle filters. A strong cooperative positioning (RCP) system has been put out to supplement GPS with ultra-wideband data in order to gain precise location (UWB). The most recent developments, however, center on visual-based localization that makes use of DL methods. This localization technique is also known as visual odometry (VO) [2].

Visual localization is possible when important landmarks in successive video frames coincide. Based on the current frame information for the automobile, salient features are provided as input to the n-point mapping technique for the automobile's position identification with reference to the preceding frame. Deep learning techniques can improve visual odometry accuracy. The accuracy of the key point detector may be impacted by these algorithms. In monocular VO, Crucial point distractors training is taught in a DNN. Computing the camera posture is another method for incrementally mapping the structure of the surroundings. The SLAM (simultaneous localization and mapping) category includes this technique [6].

SLAM is the process of simultaneously creating an online map and locating the vehicle on it. In SLAM, no a priori knowledge of the surroundings is necessary. Deep learning techniques have significantly improved picture categorization and detection; hence, these algorithms are advised as a supplement to conventional SLAM algorithms. The standard SLAM blocks might be replaced with deep learning modules to increase accuracy and robustness, even if deep learning approaches in this field are still in their infancy. AVs should be able to anticipate the motions of their surroundings to ensure safe navigation. Scene flow is the name for this. A typical method for estimating scene flow is LiDAR-based [7].

2.2.4 PERCEPTION

The use of occupancy maps for perception tasks is common. These are additionally known as the Occupancy Grid (OG). Cellular depiction of the surroundings. This technique calculates the likelihood of occupancy for each unit after partitioning the driving area into a number of units. The technique is well-established in automation and is now used as a successful solution in Automated vehicles. Deep learning algorithms are being used to follow and recognize moving objects, estimate the occupancy map surrounding the car probabilistically, and determine the context of the driving situation. When a driving scenario is derived, deep learning is used to classify the surroundings as roadway travel, parking area, or intercity travel. OG estimate relies heavily on deep learning. It helps with the image processing required to populate grid cells as well as the retrieval of features from LiDAR data. To forecast grid maps, a multitask recurrent neural network is suggested. Map layouts include drivable areas, velocity estimations, occupancy data, and semantic information [2, 8].

2.2.5 PATH PLANNING

Path planning comes next when an AV is able to locate itself in the surroundings. The capacity of autonomous vehicles to choose the best route between their starting place and their destination (i.e., the intended location) while taking into account their kinematics and dynamic model is known as path planning. The path-planning procedure should allow the automated cars to choose the optimal path to guarantee an accident-free route while keeping in mind any roadblocks it may run across in the surroundings. Since autonomous driving is a multiagent concern, the host automobile should negotiate effectively with all other road users whenever it makes a turn or changes lanes. Mission planning is the process of actively seeking the route that has been set through trajectory tracking.

Mission, motion, and behavior planning are also included in path planning. The phrase "BIG DATA" refers to the vast amount of data that is stored on the server after each time the automobile is put through a driving experience. AVs may make informed judgments in the future using the knowledge found in previously recorded data. Due to all the obstructions in the path of the vehicle, route-finding algorithms are particularly difficult. The AV should be able to recognize and steer clear of these impediments that complicate the planning algorithm's job. When faced with a particular driving circumstance or environment, the automated vehicle is aware of what should be done. For instance, a vehicle operating on a road should adhere to the planning algorithm's sequence of waypoints. For many years, researchers have studied the topic of path planning, which they frequently split into two subgroups: global route planning and local route planning. Four categories were employed to categorize the path planning approaches: Sampling, numerical optimization, interpolation, and graph search techniques [2, 8].

2.3 PLANNING TECHNIQUES

Traveling the whole state space from source location A to target location B is the basic tenet of the self-driving route planning technique. A lattice or an availability mesh are common representations of the state space, which pinpoints the positions of objects in a dynamic environment. The graph search techniques go over the availability mesh's state space and produce an ideal or substandard solution if one is discovered; otherwise, they offer no solution at all.

2.3.1 DIJKSTRA'S ALGORITHM

The proposed method is a shortest-path algorithm using a solitary resource. In this case, the phrase "Solitary-Resource" refers to the fact that there is just one available source, requiring us to find the shortest connection between the source and all of the nodes. Finding the shortest route through a mesh is done using a graph search method. It performs effectively for both organized and unstructured settings' global path planning. Despite its benefits, the method is sluggish since a lot of nodes must be visited in huge regions. Additionally, the method makes no use of any heuristics to reduce the cost of the search. Since the route was not acquired continuously, real-time scenarios cannot be used with it.

2.3.2 A STAR ALGORITHM

It is a more advanced variant of the Dijkstra method and saves time by using assumptions to ensure efficiency and speedier nodal searching. The A-star method has the advantage that the cost is determined before the node weights are established. It is good for seeking locations that the vehicle is theoretically familiar with, but it is expensive in terms of performance and storage for screening large regions. Mobile apps like any time restoring A* (ARA*) and vibrant A* (D*) use various customized versions of A-star. A* with Voronoi cost functions has been applied for path planning in parking lots and unstructured areas.

2.3.3 STATE LATTICE ALGORITHM

In the algorithm, temporal lattices are used (including velocity and time dimensions). The environment is divided into a local grid depending on the intricacy of the maneuver, making it suitable for route planning and vibrant scenarios. Regardless of its advantages, the technique is computationally costly since it must consider each conceivable solution in the database.

2.3.4 DEEP LEARNING TECHNIQUES

The two deep learning strategies used in the trajectory planning situation are planning based on reinforcement learning and mimic learning which receive the greatest attention. Imitating human driving behavior is the primary goal of imitation learning. Big data is utilized to capture the actions of the human driver, and a CNN is then used to teach the car how to strategize by copying behavior. The opposite of reinforcement learning is known as imitation learning. This technique generates driving trajectories that are identical to those of human drivers by learning how to optimize reward functions. The route is also planned to use the DRL approach. In this approach, the agent picks up driving techniques in a simulated setting.

The actual ecosystem is converted into a digital one using a transfer model as a foundation. Both of these approaches have benefits and drawbacks. Although imitation learning benefits from learned on raw figures since there is little data on corner situations, the trained network may make mistakes when handling unforeseen circumstances. On the other hand, while DRL performs well in simulations, it does not function as well in actual situations. Although deep learning-based methods for vision, localization, path planning, and control are receiving a lot of attention, the fact that deep neural networks operate in a "black box" raises questions about transparency and accountability in autonomous cars.

2.4 MOTION CONTROLLERS/ACT

The motion controller is responsible for computing steering orders (both longitudinal and lateral). To create the steering commands from sensory input, when operating as an end-to-end controller or as a component of a complete entity, the motion controller employs learning methods. The model used by conventional controllers has preset parameters. Learning controllers employ the data and training material to equip themselves with the ability to gradually learn their models. The system model becomes more precise as more data is gathered. The iterative learning control is a popular learning controller (ILC) and model predictive control (MPC).

Autonomous vehicles that follow a predefined path are examples of systems that can be effectively controlled by ILC. By resolving the optimization issue, MPC determines the proper control actions. We can foresee disturbances and uncertainties in the system with the use of MPC, which helps us arrive at the best solutions. The majority of the training data is present in the form of observations and prior states of the vehicle. Then, by teaching CNN to locate the dense occupancy grid map, it may be put to use. To determine the best trajectory for the vehicle to follow across a finite horizon, after that, the MPC cost function receives this map.

Due to the fact that these learning controllers employ both model-based control and learning algorithms, their full potential may be realized. Deep learning-based methods have become more significant in autonomous vehicle motion control. An end-to-end convolutional neural network model (from visuals to control instructions) is trained using a visual attention model. The picture areas impacting the output of the network are identified by these attentions, which the attention model has learned. A video-to-text approach based on attention is utilized to provide textual explanations. Finally, the explanations are grounded in the visual regions that were important to the controller by aligning with his or her attention map. Perception, judgment, and control systems are the three fundamental parts of autonomous cars, and they have all been thoroughly studied separately. However, it is also possible to use DNN's strength to improve sensing, decision-making, and motion control jointly [2, 3].

2.5 CONCLUSION

AI is a key factor in the advancement of vehicle technology in the automotive sector. Mechatronics and AI work together to ensure that vehicle operations are carried out correctly. The Global Positioning System (GPS), onboard sensors including laser, radar, and lidar, as well as vehicular communication networks, are used by autonomous cars to gather scene information. The information collected is then put to use in a variety of path planning and control strategies to equip the vehicles with the ability to operate independently in challenging terrain. Modern AI algorithms are used by autonomous cars to locate themselves in both known and unknown surroundings. For vision, path planning, and motion control, AI algorithms are also used [2].

KEYWORDS

- **artificial intelligence**
- **automotive sector**
- **autonomous vehicle**
- **deep learning**
- **mechatronics**
- **navigation architecture**
- **path planning**

REFERENCES

1. Qing, L., Da-Chuan, L., Qin-Fan, W., Liang-Wen, T., Yan, H., Yi-Xuan, Z., & Nong, C., (2013). Autonomous navigation and environment modeling for MAVs in 3-D enclosed industrial environments. *Computer in Industry*.
2. Neelma, N., Muhammad, K. E., Muhammad, R. A., Md Yeakub, A., & Muhammad, A. Q., (2022). Intelligence of autonomous vehicles: A concise revisit. *Journal of Sensors*.
3. Gang, C., Thierry, F., & Luis, M., (2007). A real-time autonomous navigation architecture. *IFAC Proceedings Volumes*.

4. Jesús, M., Jorge, L. M., María, A. M., & Anthony, M., (2009). Pure-pursuit reactive path tracking for nonholonomic mobile robots with a 2D laser scanner. *EURASIP Journal on Advances in Signal Processing.*
5. António, S., Urbano, J. N., & Rui, A., (2005). Fusion of odometry with magnetic sensors using Kalman filters and augmented system models for mobile robot navigation. *IEEE International Symposium on Industrial Electronics (ISIE 2005).*
6. Hügli, H., Facchinetti, C., Tièche, F., Müller, J. P., Rodriguez, M., & Gat, Y. (1994). Architecture of an autonomous system: application to mobile robot navigation. In *Proceedings of Symposium on Artificial Intelligence and Robotics* (pp. 97–110).
7. Gutiérrez, R., López-Guillén, E., Bergasa, L. M., Barea, R., Pérez, Ó., Gómez-Huélamo, C., ... & López-Fernández, J. (2020). A waypoint tracking controller for autonomous road vehicles using ROS framework. *Sensors, 20*(14), 4062.
8. Katal, A., & Singh, N., (2022). Artificial neural network: Models, applications, and challenges. In: Tomar, R., Hina, M. D., Zitouni, R., & Ramdane-Cherif, A., (eds.), *Innovative Trends in Computational Intelligence.* EAI/Springer Innovations in Communication and Computing. Springer, Cham. https://doi.org/10.1007/978-3-030-78284-9_11.
9. Singh, N., Prateek, M., & Chauhan, P., (2019). An efficient path planning algorithm for networked robots using modified optimization algorithm. In: *International Journal of Innovative Technology and Exploring Engineering* (Vol. 8, No.12, pp. 2738–2743). Blue Eyes Intelligence Engineering and Sciences Engineering and Sciences Publication – BEIESP.
10. Singh, N., Prateek, M., Chauhan, P., & Choudhury, T., (2020). An optimized self-learning algorithm for autonomous navigation in an unknown environment. *Journal of Green Engineering, 10,* 3315–3334.

CHAPTER 3

Algorithms for Robot Positioning and Navigation

KINGSLEY IGULU,[1] WILSON NWANKWO,[2] JUSTICE PALIMOTE,[3] and FRIDAY ONUODU[4]

[1]*Department of Computer Science, Ignatius Ajuru University of Education, Port Harcourt, Nigeria*

[2]*Department of Computer Science, Edo State University, Uzairue, Nigeria*

[3]*Department of Computer Science, Rivers State Polytechnic, Bori, Nigeria*

[4]*Department of Computer Science, University of Port Harcourt, Port Harcourt, Nigeria*

ABSTRACT

A robot is a machine that is situated in some environment (experiences environment via sensors and acts via actuators), capable of autonomous (control over state and behavior) actions in the environment so as to fulfill the design objectives or intentions (reactivity and proactivity). Depending on the application area and the objectives of a robot, a robot may require navigation to fulfill its goals. Applications for autonomous robots range from the military to transportation to industry to space exploration. Material handling, disaster relief, patrolling, and rescue operations are all tasks that mobile robots are capable of performing. Because of this, an autonomous robot is needed that is capable of moving freely in both a static and

Advances in Autonomous Navigation Through Intelligent Technologies.
Niharika Singh, Thipendra P. Singh, Brian Azzopardi (Eds.)

dynamic setting. The primary goal of mobile robot navigation is to ensure safe and efficient movement of the robot from one point to another in a cluttered environment while maintaining a clear path and generating the shortest possible path. Environment mapping, trajectory design, motion control, and obstacle avoidance are all part of an autonomous mobile robot's navigation process. In order to perform well in navigation, it is necessary to have both global and local navigation. Positioning (localization) and navigation in robotics are crucial aspects that have attracted the attention of researchers. This chapter will present different research works on robot positioning and navigation. Mobile robots use numerous types of navigation and positioning, including visual navigation and positioning, Global Positioning System (GPS) global positioning, electromagnetic navigation and positioning, and light reflection navigation and positioning. This chapter will hopefully shed light on the numerous algorithms utilized for positioning and navigating mobile robots.

3.1 INTRODUCTION

Depending on the application area and the objectives of a robot, a robot may need navigation to achieve its goals. Autonomous robots have a wide range of applications, from military to transportation, industry, and space exploration. Tasks such as material handling, disaster relief, patrolling, and rescue operations can all be carried out by mobile robots. Therefore, there is a need for an autonomous robot that can move freely in both static and dynamic settings. This means that we will have to restrict our focus to mobile robots. The primary goal of mobile robot navigation is to guarantee the robot's safe and effective movement from one location to another within a cluttered environment, all the while keeping the robot on a clear path and generating the shortest path possible. The navigation process of an autonomous mobile robot includes environment mapping, trajectory design, motion control, and obstacle avoidance. Both global and local navigation skills are required for optimal performance.

During navigation, the robot makes use of a suite of high-tech instruments to construct a mental representation of its surrounding area, pinpoint its precise location, regulate its speed and direction, identify potential dangers, and plot a course around them. Planning a secure path between the origin and destination is the primary purpose of any navigational method. Choosing the right navigation method is crucial when plotting a course for a robot in either a straightforward or intricate setting. One of the most actively researched facets of robotics right now is navigation for mobile robots. Different branches of mobile robot navigation include in-robot navigation, global navigation, and local navigation. A robot's ability to define its position in space with respect to a reference axis and then navigate to a specified destination is called global navigation. When performing local navigation, however, the mobile robot does not need to be familiar with its surroundings in order to avoid obstacles and reach its destination. Understanding the environment's changing dynamics and establishing spatial relationships among its various elements are two of local navigation's most important goals. A local navigation strategy is used when there is only a partial or complete understanding of the surroundings. Personal navigation is the process of navigating one's surroundings by paying attention to the relationships between the various elements [1]. A flowchart of how a robot works can be seen in Figure 3.1.

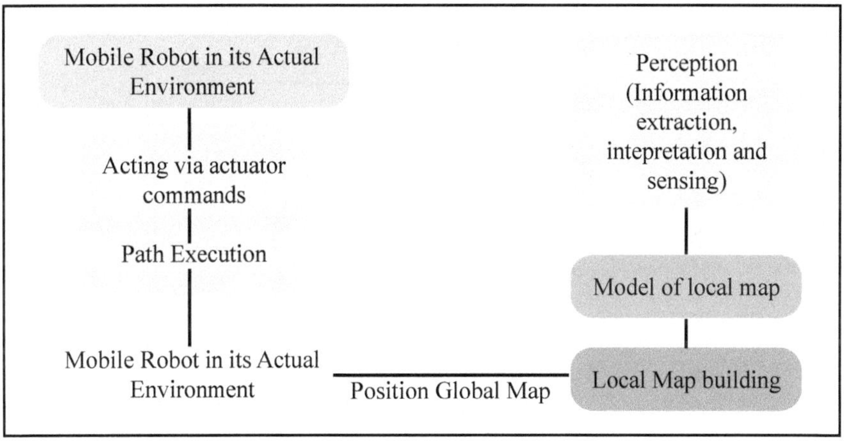

FIGURE 3.1 Functioning of General Robot Navigation.

Source: Modified from Ref.[1]

The capability to navigate the environment is one of the most difficult skills a mobile robot must possess. The robot must excel at perceiving its environment, determining its precise location within that environment, processing pertinent data, and controlling its movement for navigational success. The robot's ability to perceive its environment depends on its ability to process the information it receives from its sensors and draw conclusions regarding the actions it must take to achieve its objectives [2]. In this chapter, we shall collapse the components of navigation into localization and path finding. In other words, we shall explore various localization and path finding algorithms.

3.2 LOCALIZATIONS AND POSITIONING

A robot's current location can be determined using localization techniques. "Where am I?" is answered in a nutshell by localization. To be truly localized, a robot must do more than simply be aware of its exact location in relation to the Earth's axis (that would mean positioning alone). It suffices to note that positioning is an integral part of localization. Imagining a robot that interacts with humans as an example of this. This robot may need to know its exact location, but it also needs to know where it stands in relation to its human targets. Using its sensor array, it can identify humans and then calculate its position in relation to them. Mobile robot localization is depicted in Figure 3.2. We first present various localization technologies in subsequent subsections.

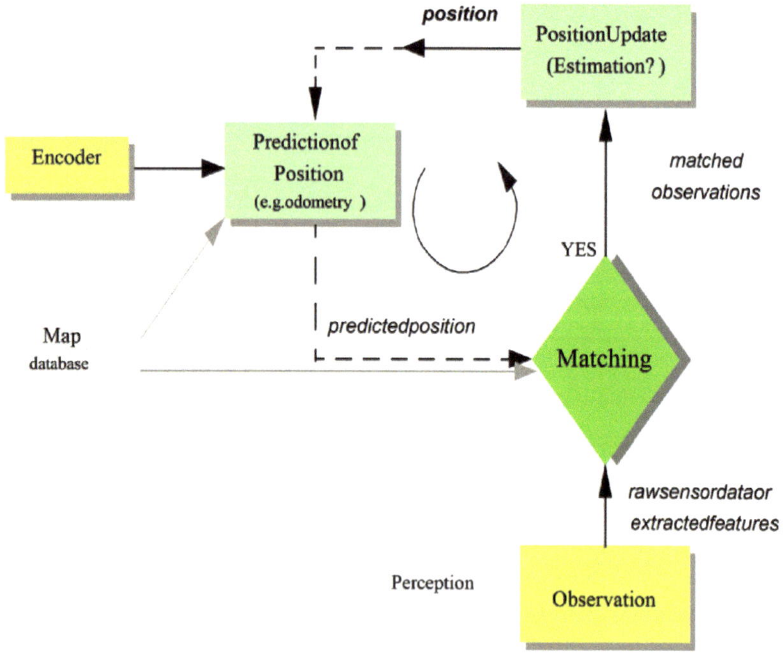

FIGURE 3.2 General schematic for mobile robot localization.

Source: Reprinted with permission from Ref.[2]

3.2.1 ODOMETRY

Odometry is the most popular navigation method for mobile robot positioning due to its short-term accuracy, low cost, and ability to sample at a high rate. However, odometry is dependent on the incremental accumulation of motion data over time, which results in an unbounded increase in error. Large lateral position errors, caused in part by improper robot orientation, tend to amplify as the robot moves further from the target. Researchers concur that odometry is an essential component of robot navigation systems, and that if its precision could be enhanced, robots would have an easier time navigating their environment. According to Chenavier & Crowley [3] and Cox [4], it is possible to combine odometer data with absolute position measurements for a more precise estimate.

As long as one can reliably translate wheel revolutions into linear displacement with respect to the ground, the equations of odometry [5] hold. Wheel slippage and other subtler causes show that not all rotational motion is linear. Systematic errors and non-systematic errors are the two types of errors that can be found in the results. It is common for robot-related kinematic errors, such as unequal diameter wheels or a nebulous wheelbase, to lead to systematic errors. For example, bumps and cracks on the floor can lead to non-systematic errors such as wheel slippage. Landmark density is typically determined empirically, and the worst-case systematic errors are taken into account when setting the density of landmarks for an odometry/landmark navigation system. Large non-systematic errors are likely to bring down such systems.

3.2.2 INERTIAL NAVIGATION

When adopting inertial navigation, gyroscopes, and accelerometers are used to track the rate of rotation and acceleration, respectively. Calculating location is as simple as multiplying measurements by two (or more in the case of accelerometers). Due to the fact that they are self-contained, inertial navigation systems eliminate the need for outside references. Due to the need to integrate rate data to obtain position, inertial sensor position data drifts over time, and any small constant error grows without bound after integration. Therefore, inertial sensors are not ideal for precise, permanent positioning.

3.2.3 MAGNETIC COMPASS

In terms of total dead reckoning errors, vehicle heading is the most crucial of the three navigation parameters (x, y, z). As a result, absolute heading sensors are critical for autonomous platforms in solving their navigation issues. This is the case with sensors like the magnetic compass. In urban areas, where there are many steel buildings and power lines, the Earth's magnetic field is distorted, making it difficult to use a magnetic compass. This makes it challenging to implement geomagnetic sensors in indoor settings. There are a number of sensor systems out there that rely on different physical effects related to the Earth's magnetic field in order to collect data, namely:

i. Hall-effect compasses;
ii. Mechanical magnetic compasses;
iii. Magneto-resistive compasses;
iv. Magneto elastic compasses;
v. Fluxgate compasses.

When constructing a mobile robot, the fluxgate compass is the optimal choice. Due to its low power consumption, lack of moving parts, shock and vibration resistance, rapid startup, and low cost, fluxgate compasses are well-suited for measuring the horizontal magnetic field of the Earth. If the vehicle is designed to traverse difficult terrain, the vertical component of the geomagnetic field can cause significant errors. To prevent this, the sensor coil should be gimbal-mounted and mechanically damped.

3.2.4 ACTIVE BEACON

Active beacon navigation systems are utilized by all ships, aircraft, and commercial mobile robots. Accurate positioning information can be obtained with minimal processing from active beacons. As a result, high sampling rates and high reliability can be achieved, but the installation and maintenance costs can be quite high. Beacons must be mounted precisely to ensure accurate positioning. Trilateration and triangulation are two distinct types of active beacon systems.

In trilateration, the distances between a moving vehicle and three fixed points, or beacons, are measured in order to pinpoint its exact location. Three or more transmitters fixed in known locations around the environment communicate with a single receiver on the robot itself; this configuration is called a trilateration navigation system. On the flip side, it's possible that there's only one transmitter and all receivers are wall-mounted. Time-of-flight data is used to determine the physical separation between the fixed transmitters and the mobile receiver. Trilateration can be seen in action in Global Positioning Systems (GPS).

The use of triangulation with active beacons is a popular, accurate, and flexible absolute localization method. Several well-known triangulation algorithms have been discussed in the academic literature, including triangulation with two circle intersection, geometric triangulation, Newton-Raphson iterative search, iterative search, geometric circle intersection, the algorithm from the imperial college beacon navigation system, triangulation with three circle intersection, and the position estimator algorithm [6].

3.2.5 GLOBAL POSITIONING SYSTEMS (GPS)

As far as outdoor navigation goes, the Global Positioning System (GPS) is an absolute game-changer. When GPS was first conceived, it was intended to be a Joint Services Program within the Department of Defense. To send encoded radio frequency (RF) signals, the system uses a total of 24 satellites, plus 3 backups. Radio frequency (RF) signals sent from satellites to ground-based receivers can be used by the latter to triangulate their precise location on the planet. It is possible for receivers to determine the satellites' locations by measuring the time it takes for these signals to travel, as they contain information about the satellites' current locations. In theory, a ground receiver's latitude, longitude, and altitude could be determined given only the distance between the receiver and the three satellites [1].

3.2.6 LANDMARK NAVIGATION

The sensors on a robot will be able to pick out landmarks because of their one-of-a-kind characteristics. There is no restriction on the types of geometric shapes that can serve as landmarks; examples include rectangles, lines, and circles. These landmarks can even contain supplemental data (e.g., in the form of barcodes). In most cases, landmarks can be used by a robot for localization because of their known and consistent position. In order for landmarks to be easily identifiable, they are meticulously selected based on criteria such as their contrast to the surrounding environment. If a robot does not already have the relevant information about landmarks stored in its memory, it will be unable to use them as a means of navigation. The primary challenge of localization is thus to accurately identify landmarks and determine the robot's position.

It is a common practice to make the assumption that the current position and orientation of the robot are roughly known in order to simplify the process of landmark acquisition. This allows the robot to search for landmarks in a more condensed area. Because of this, precise odometry is absolutely necessary in order to locate landmarks.

There are two main types of landmarks: those that have existed since the beginning of time (natural) and those that were created by humans (man-made). Just keep in mind that "natural" landmarks work best in highly structured environments like factories, hospitals, and hallways. A natural landmark is an object or feature already present in the environment that serves a function other than guiding robots through their immediate vicinity. In contrast, artificial landmarks are man-made objects or markers strategically placed in the environment to aid robots in navigating their respective regions [1].

3.2.7 MAP-BASED POSITIONING

Map-based positioning, also known as "map matching," is a method in which the robot uses its sensors to create a three-dimensional representation of its immediate environment. Then, this regional map is compared to an existing mental representation of the world. A robot's ability to pinpoint its location and orientation depends on whether or not it has detected a suitable match. The stored map may be a 3D CAD model of the area. It may also be reconstructed from historical sensor data. Unlike other methods of indoor positioning, map-based positioning doesn't require any alterations to the built environment in order to provide accurate location data. With the help of some of the emerging algorithms, map-based positioning also enables a robot to acquire knowledge of its surroundings in order to refine its positioning as it explores. The main drawbacks of map-based positioning are the need for a large number of fixed, easily distinguishable features and the high accuracy of the sensor map. As a result of these stringent conditions, most map-based positioning research is still conducted in controlled lab settings or in simplified real-world settings.

3.3 NAVIGATION ALGORITHMS

The navigational algorithms used by mobile robots can be divided into two categories: the classical and the reactive. Roadmap approach (RA), cell decomposition, and artificial potential field are some examples of classical algorithms. However, this list is not exhaustive. The following are examples of reactive techniques: artificial bee colony (ABC), genetic algorithm (GA), neural network (NN), fuzzy logic (FL), firefly algorithm (FA), ant colony optimization (ACO), bacterial foraging optimization (BFO), particle swarm optimization (PSO), shuffled frog leaping algorithm (SFLA), cuckoo search (CS), and other miscellaneous algorithms (OMA) (Figure 3.3).

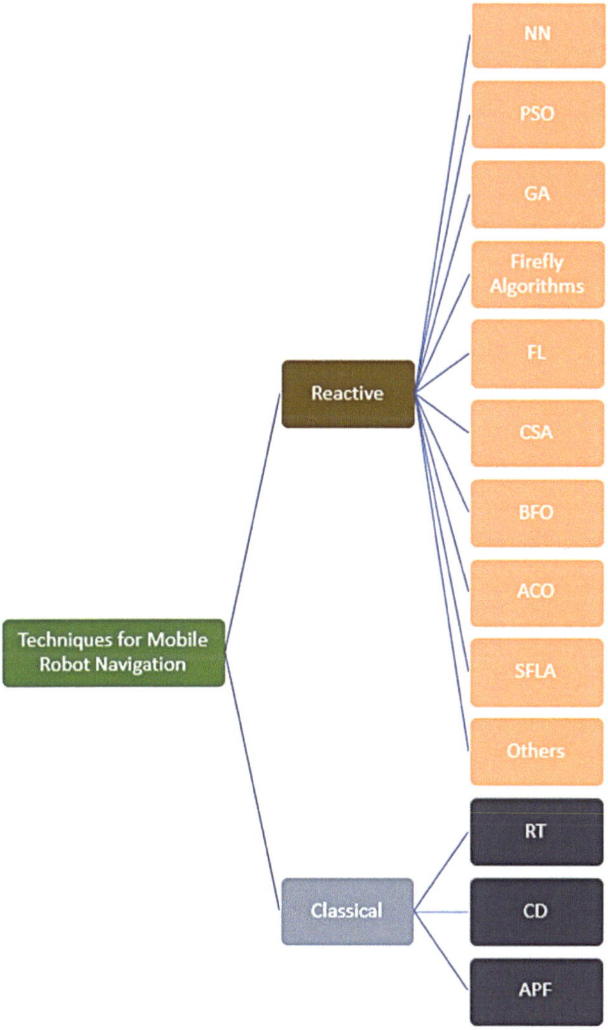

FIGURE 3.3 Classification of mobile robot navigation algorithms.

3.3.1 CELL DECOMPOSITION NAVIGATION

This strategy achieves its objectives by segmenting the terrain into grids (cells) that are not connected to one another and relying on connectivity graphs to direct pedestrian traffic between the cells. Pure cells, which are cells without obstructions, are taken into account when planning a route from the starting position to the desired location. Along the way, cells that are corrupted (cells that contain obstacles) are split in half to create a new, clean cell, which is then used to determine the most direct route from the origin to the destination. This guarantees the quickest route is taken. After that, the next uncontaminated cell is added. Beginning and ending cells are "stand-ins" for the origin and destination of the CD procedure [7, 8]. This route is being taken, as shown by the continuous string of clean cells between those coordinates. An individual can classify the CD approach as adaptive, approximate, or exact. Rather than having a predetermined form, cells in the exact cell decomposition [9] take on the form of their surroundings, including the size and placement of any obstacles. Different applications of the regular grid are used in this tactic. After giving each component a number, the initial unoccupied area of the environment is partitioned into smaller components (trapezoidal and triangular). Each part of the environment can be thought of as a node in a connectivity graph. Following this, connections between neighboring configuration nodes are allowed, and a path in this chart is like a diversion in free space, represented by the sequence of striped cells. Connecting the series of striped cells represents a network in this diagram. The crossing points of adjacent cells in the channel are then used to link the underlying arrangement to the objective design, thereby transforming the channel into a free path. This creates a bridge between the two concepts.

3.3.2 ROADMAP APPROACH (RA)

RA is commonly known as the highway approach. It is the path taken from one location to the other, and it is represented by a series of one-dimensional curves connecting the numerous open areas [10]. The full road map is a compilation of comparable routes that the planner can use to zero in on the optimal arrangement. Nodes play a crucial role in this case in determining the best route for the robot to take. The RA calculates the quickest route from the robot's current location to the target using Voronoi and visibility graphs. The beginning and ending locations are linked using map nodes in the visibility graph method. In Figure 3.4, we see a representation of the visibility graph, where the shaded area indicates obstructions, and the dashed line represents the path from A to B.

The nodes in a scenario where obstacles are polygons are the vertices, and the edges are the lines between them [11]. Alternative methods for robot path planning employ the Voronoi diagram [12]. Using this strategy, the area is segmented, and figures are pieced together by linking points that are uniformly spaced around the boundary of the challenge.

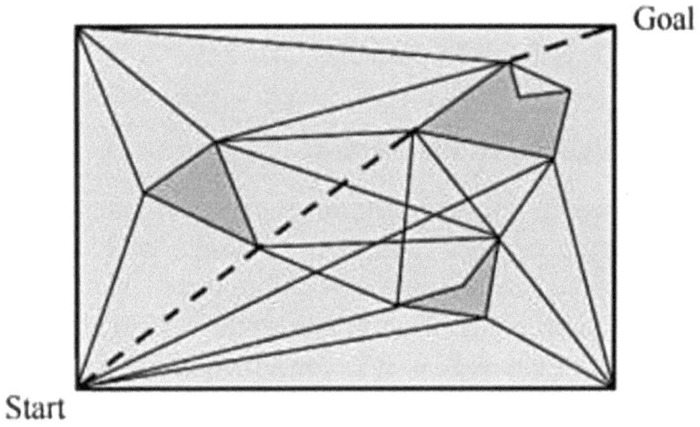

FIGURE 3.4 Visibility graph.

Source: Reprinted with permission from Ref.[13]

3.3.3 APPROACH USING AN ARTIFICIAL POTENTIAL FIELD (APF)

The APF method was first presented by Khatib [14] for controlling mobile robots. The "imaginary force" of the robot, he says, is the sum of the potential of the goal and the obstructions, which function as charged surfaces. The robot is being guided to its destination by an imaginary force that also keeps it from colliding with an obstacle. In this case, the robot follows the negative gradient to sidestep the obstacle and get where it needs to go. Garibotto & Masciangelo [15] implemented this method for autonomous robot movement. Using APF, Kim & Khosla [16] describe a novel approach to navigating hazardous environments. A harmonic function was used to get around the local minimum issue. Also, Borenstein & Koren [17] have proposed a way to deal with the issue of local minima. The dynamical aspects of robot navigation have been taken into account in this analysis. This chapter examines how well APF works in a dynamic setting to avoid obstacles [18]. The APF method can be modified using electrostatic principles [19]. By employing electrostatics, we can construct the potential function and determine the path free of collisions in real time. Although it is challenging to avoid obstacles in real time, Huang [20] developed a velocity controlling mechanism to detect and avoid them. To help solve optimization problems, avoid local minima, and find global optima, Shi & Zhao [21] introduced the superior potential function and the superior repulsive potential function. Sfeir et al. [22] discuss the use of APF strategies like oscillation and conflicts in the context of mobile robot navigation. They reworked the APF to reduce oscillation and conflicts when the target is close to an obstacle.

3.3.4 GENETIC ALGORITHM (GA)

Discovered by using the principles of genetics and natural selection, this search-based optimization technique has become widely used [23]. The first use of it in computer science was introduced in 1975 [24]. It has since found widespread application across the STEM disciplines and is even used for robot navigation. It is concerned with the maximization or minimization of an objective function subject to constraints. Using this strategy, the population (a group of individuals who can be distinguished from one another based on their genetic makeup) is assigned to the problem at hand, and fitness scores are calculated for each individual based on how well they meet the goal. These individuals have been selected for their fitness and will be used in a genetic crossover to pass on their traits to future generations. In doing so, the mutation protects the unique characteristics of each population member and delays the onset of uniformity. Once the population converges, the algorithm stops. The GA may be somewhat haphazard, but it outperforms a random local search because it can potentially draw on past data.

Shing & Parker [25] introduced a real-time path planner in response to the extreme latency of traditional search and optimization techniques. GA is a powerful search technique that can find information quickly in an environment you don't know much about. This strategy is employed by Xiao et al. to accomplish navigational objectives like path length, path smoothness, and obstacle avoidance. Kang et al. [26] approach the problem of getting lost in unfamiliar territory from a non-linear environmental perspective. They have suggested an online training model for obtaining the fittest chromosomes in order to avoid getting stuck and find a way out of tricky situations. Shi & Cui [27] propose using GA to navigate past a shifting barrier in an uncharted environment, whereas the majority of studies on GA for navigation have concentrated on a fixed environment. Many researchers have combined GA with another intelligent algorithm to create hybrid approaches, such as GA-FL [28], GA-NN [29], and GA-PSO, to enhance GA's already impressive performance in robot path planning [30]. The task of path planning for a group of robots is one of the difficulties faced by robotics. Kala [31] used GA to investigate the path planning strategy of numerous mobile robots. In order to decrease the amount of robot collisions, his research focused on creating a method to effectively coordinate the movements of multiple robots in a static environment. Liu et al. [32] demonstrate a technique for multiple objectives in a static environment that is very similar to multiple robot path planning. Numerous studies have shown that the GA has a number of flaws, including a slow convergence rate, no assurance that the best solution will be found, a laborious process for figuring out the parameters for mutation rate and population size, etc.

3.3.5 FUZZY LOGIC (FL)

After being presented for the first time by Zadeh in 1965 [33], the idea of FL has since been adopted in a wide variety of fields of scientific investigation and technological advancement. It is used in situations where there are a significant number of unknowns, along with complexity and nonlinearity. This category encompasses a wide range of activities, including pattern recognition, autonomous control, decision making, and data classification. The FL framework concept receives additional support from the extraordinary capacity of human beings to process perceptual data. To be more specific, it develops mathematical models based on the rules (If-Then statements) supplied by humans. This makes the life of the designer and the computer easier by providing more precise data on how systems perform in the real world. This is used for mobile robot path planning to make the designer's job easier. Zavlangas & Tzafestas [34] provide an explanation of fuzzy (Sugeno) navigation for an all-seeing mobile robot. The research carried out by Castellano et al. came up with the autonomous fuzzy rule generating method for obstacle avoidance for efficient navigation. This method was developed by them in 1997. FL describes the navigation system in an unstructured static and dynamic environment, which enables it to avoid common navigational pitfalls such as endless looping, backtracking, and dead-end traps (U-shaped, maze, snails), as well as steering from restricted passages and curved trajectories. FL is used in conjunction with a wide variety of navigation techniques these days. These techniques include sensor-based methods for better incremental learning of the new environment, reinforcement-based methods for reducing angular and radial uncertainties in the environment, and algorithm-based methods like NN [35], GA [36], APF [37], ACO [38].

Data-driven FL was used by both Al-Khatib & Saade [39] and Lee et al. [40] to overcome navigational challenges in a dynamic environment. Cooperative strategies are recommended for the navigation of multiple mobile robots in an unfamiliar and crowded environment [41]. Kang et al. [42]; Al-Mutib et al. [43]; and Kang et al. [92] present a stereovision-based mechanism that employs FL to enhance the robot's performance in a dynamic setting. To keep tabs on a moving object, Abadi & Khooban [44] developed a Mamdani-based FL controller for a wheeled mobile robot. PSO and FL were combined to form a hybrid strategy for determining optimal parameter values. In order to maintain diversity control in ACO and forestall early convergence, Castillo et al. [45] found that FL could effectively operate. With the help of Al-Jarrah and colleagues, a strategy was developed wherein one robot acts as a leader and other acts as a follower. While a low-level controller was implemented in the companion robot, a high-level controller based on the first order Sugeno fuzzy system was developed for the lead robot. The learning strategy is developed using NN, and efficient fuzzy rules are optimized using ANFIS. To better direct and guide missiles, UAVs, and ROVs, FL has been implemented in the defense technology sector.

3.3.6 ARTIFICIAL NEURAL NETWORK (ANN)

The artificial neural network is a very complicated information processing system that, when broken down into its component parts, is actually made up of a great number of straightforward elements. This information is transmitted through the dynamic responses of the elements to the stimuli that come from the environment. The NN is represented by layers of neatly arranged nodes that are connected to one another. Every single one of the nodes has an activation function that was constructed right into it. The input layer of the NN mechanism is responsible for recognizing the patterns. A network of weighted connections is used to facilitate communication between these patterns and the unseen layers that are being processed. Due to the fact that the output layer is connected to the hidden layers, it is able to produce the required response. Because of its superior performance in areas such as generalization, massive parallelism, distributed representation, learning, and fault tolerance, neural networks (NN) have emerged as a compelling candidate for mobile robot navigation. Janglová provided an explanation in 2004 of how a neural network could assist a mobile robot on wheels in navigating an unfamiliar environment. In order to produce a route free of obstructions, he made use of two procedures that are based on neural networks. The first neural mechanism gathers data from the senses in order to locate an open area, and the second neural network locates a route that is risk-free by avoiding the obstacle that is located closest to it. Qiao et al. [46] proposed a method for automating the learning process as a means of removing the necessity of relying on human guidance throughout the process of navigation. The novel aspect of their work is that the neural network automatically adjusts the number of hidden layers it trains with by either adding to or removing them depending on the degree to which the surrounding environment challenges it. The addition of neural networks (NN) to the Fast Simultaneous Localization and Mapping (Fast SLAM) method was suggested by Li et al. [47] as a way to reduce the number of errors caused by an inaccurate odometry model and a linearization of the SLAM nonlinear function. This was done in order to make the method more accurate. When NN is combined with Fast SLAM, the mobile robot's capability to navigate an unknown environment without colliding with obstacles is improved. In order to achieve the most successful outcomes possible for robot navigation, a number of different approaches, one of which is NN, are combined into a hybrid mechanism. Na & Oh [48] demonstrated that behavior-based control can be coordinated both cooperatively and competitively by combining the NN and APF. This was accomplished by combining the two models.

3.3.7 PARTICLE SWARM OPTIMIZATION (PSO)

Particle swarm optimization (PSO) was initially developed by Eberhart & Kennedy in 1995 [49] as an optimization tool for the purpose of solving difficult problems in the fields of engineering and the natural sciences.

PSOs behave in a manner that is comparable to that of social animals, but they do not require the presence of a designated leader in order to complete their tasks. It is not necessary for a bird flock to have a leader in order to locate the best food because the birds will simply follow whoever is in the neighborhood (Figure 3.5). Because of this, the birds are able to successfully communicate with the community in order to obtain the necessary solution. An alternative solution to the issue at hand is represented by each individual particle that makes up the PSO algorithm. PSO has seen increased use in recent times as an approach to the challenge of mobile robot navigation. The difficulties of mobile robot navigation in an unknown environment were addressed by Tang et al. [50] by employing a multi-agent particle filter to solve the problems of mapping and localization. This was done in order to address the difficulties of mobile robot navigation in an unknown environment.

The use of PSO helps to speed up computations and maintain stable convergence characteristics. Ha et al. [51] have combined the PSO and MADS algorithms in order to obtain a precise trajectory and prevent themselves from becoming mired in local optima (Mesh Adaptive Direct Search). When the PSO MADS algorithm is applied, both the GA and the EKF show signs of improvement (Extended Kalman Filter). Atyabi et al. [52] developed the Area Extended PSO (AEPSO) as an extension of the original PSO to address the time- and space-varying constraints that are inherent in mobile robot navigation. This was done by extending the original PSO. Bomb disposal and other types of rescue operations have both found successful application of the AEPSO method. Tang & Eberhard [53] have requested assistance from the PSO in order to coordinate the movement of multiple robots through an intricate environment. The proposed method, which utilizes the PSO in conjunction with the multibody system dynamics, which includes acceleration, mass, force, and inertia similar to those of a robot, is then subjected to a fault tolerance analysis. This analysis is carried out after the proposed method has been implemented. The real-world navigation of a number of mobile robots has been improved thanks to the work of Couceiro et al. [54]. They adapted PSO and Darwinian PSO in order to address concerns regarding safety and coordination (DPSO). In a system consisting of 12 physical robots, they discovered that the maximum communication distance was as high as 90%, and that the global optimal efficiency was even higher. Chen et al. [93] made use of a multi-category classifier in their attempt to design a human expert control approach with learning-based capability for an uncertain environment. This was accomplished by classifying variables into multiple categories. As a direct consequence of this, the PSO is utilized in order to quicken the process while preserving a high level of accuracy. A standard grid search cannot compare to the accuracy of this one.

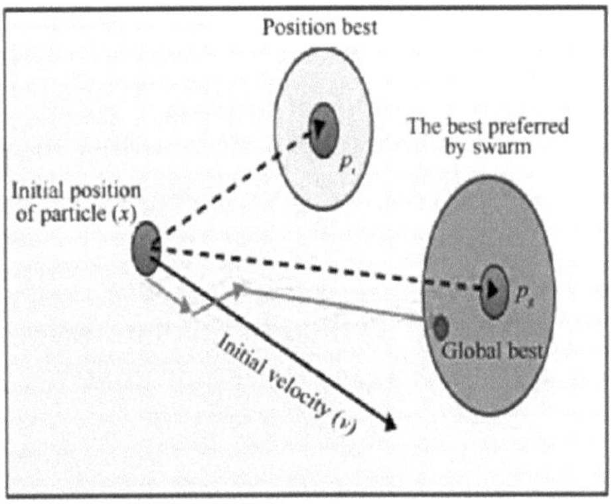


```
PSO pseudo-code
while maximum iteration or minimum error criteria
is not attained do
    for each particle do
        initialize particle
    end for
    for each particle do
        Calculate the fitness value
            If the fitness value is better than the best fitness
value in history (pbest)
                Set current value the new pbest
                end if
            end for
for each particle do
    Find in the particle neighborhood the particle with the
best fitness (gbest)
    Calculate particle velocity using velocity equation
    Apply the velocity construction
    Update the particle position using position equation
    Apply the position constriction
end for
end while
```

FIGURE 3.5 Basics of PSO and pseudocode.

Source: Reprinted with permission from Ref [13]

3.3.8 ANT COLONY OPTIMIZATION (ACO)

An algorithm for swarm intelligence was initially developed by Marco Dorigo in 1992 [94] as part of his doctoral dissertation [55]. A population-based method is utilized in this method of combinatorial optimization, which gives the approach its name. The ACO algorithm is designed to function in a manner analogous to the foraging behavior of ants by locating the shortest route from the colony to a source of food (Figure 3.6). The ACO algorithm has been put to use in a wide variety of scientific and engineering domains, including the scheduling of job shops, the routing of vehicles, the quadratic assignment problem, the traveling salesman problem, and graph coloring. The ACO is currently being applied to the problem of mobile robot navigation in order to minimize the risk of collision with obstacles and maximize the effectiveness of route planning. Tan et al. [56] detailed the use of ACO for real-time route planning for mobile robots in their research. The ACO is superior to the GA in a number of respects, such as the speed of convergence, the variation of solutions, the efficiency with which computations are performed, and the dynamic convergence behavior. The purpose of this chapter is to present an ACO-based navigation system for use with a group of mobile robots [57]. They devised a system that allowed multiple robot systems operating in the same environment to avoid colliding with one another. As a consequence of this, they were successful in improving the selectivity of the strategy by making use of a novel function. A penalty function is applied to the trail intensity whenever an ant reaches a dead end in order to prevent the path taken by the robot from becoming impossible to navigate. Castillo et al. describe a hybrid ACO-fuzzy approach to mobile robot navigation. This approach can be found in the literature [45]. Kumar et al. [58] describe a RA-ACO-based strategy that can be used by humanoid robots for navigation in crowded environments. They found that the simulated and actual results had an excellent level of agreement after putting the proposed method for humanoid robot navigation through its paces in a real-time Petri-Net environment and finding that it performed admirably.

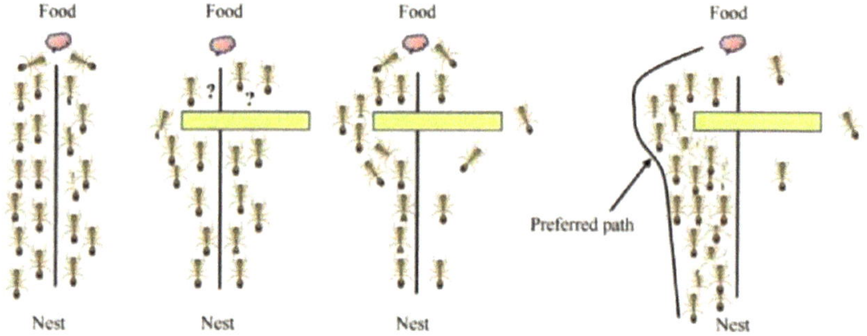

FIGURE 3.6 Ants' behaviors while searching for food.

3.3.9 BACTERIAL FORAGING OPTIMIZATION (BFO) ALGORITHM

In 2002, Passino [59] presented an innovative method for optimizing data that was inspired by the work done by the bacteria *E. coli* and *M. xanthus*. These bacteria look for food in the most effective way possible using a variety of different strategies. In the process of chemotaxis, also known as the detection of chemical gradients, which enables bacteria to exchange specialized signals, the BFO algorithm is utilized. It is supported by the foundation of four fundamental concepts, which are chemotaxis, swarming, reproduction, and extinction, as well as spread. The activities of the bacteria as they search for a source of nourishment [60].

Bacteria are in a state of constant motion when it comes to food. Bacteria that are located in areas with an abundance of nutrients flourish and grow by a factor of two, whereas bacteria that are located in areas with a deficiency of nutrients either disperse or perish. Bacteria that are located in an area with an abundance of nutrients emit a specific warning signal to other bacteria, whereas bacteria that are located in an area with a deficiency of nutrients are chemically drawn to one another. There is a very favorable environment for the growth of bacteria here. Bacteria have once more started to disperse across the globe in the hopes of finding a new region that is abundant in nutrients.

Sierakowski & Coelho [61] were the first people to apply the BFO algorithm to the problem of mobile robot navigation in a static environment with variable velocity. This environment followed a uniform distribution as well as the Gaussian and Cauchy distributions. Gasparri & Prosperi [62] provide real-time navigation in corridor, lobby, and building floor environments for an individual mobile robot system. Abbas & Ali [63] develop a brand new BFO algorithm with the goal of improving the performance of the path planning for a wheeled robot. The newly developed method makes use of an APF technique to model the environment with two opposing forces, specifically an attractive force for the goal and a repulsive force for the obstacles; the method then analyzes the algorithm's negative feedback to select appropriate direction vectors that lead the search process to the promising area with a more efficient local search. Jati et al. [64] propose a BFO algorithm as a potential solution to the difficult problem of navigation when there are a large number of robots present. The harmony search algorithm and the BFO were both incorporated into their project by the group. In addition to the wheeled robot, the BFO algorithm has been tested and shown to be effective for an industrial manipulator. Oyekan & Hu [65], in their 2010 study, addressed the problem of UAV navigation utilizing BFO. A proportional integral derivative (PID) controller is implemented in the BFO in order to obtain optimal search parameters in 3D space and to avoid complex modeling when tuning the controller for an unmanned aerial vehicle. This is done in order to maximize efficiency (UAV).

3.3.10 ALGORITHM FOR ARTIFICIAL BEE COLONY (ABC)

Figure 3.7 is a representation of the honeybees' foraging activities, which served as the inspiration for Karaboga's [67] proposal of the ABC algorithm, which is an intelligent swarm-based technique. The ABC algorithm is an example of a population-based algorithm; in order to solve problems, it draws from a pool of previously provided answers (food source for bees). Because it is a swarm algorithm, we can deduce that it is a population-based stochastic search method that features an intuitive user interface and blazing-fast throughput. The food search cycle of ABC consists of the following steps: (i) send worker bees to a potential food source so that they can evaluate the quality of the nectar there; (ii) observers select potential food sources based on the data gathered from the worker bees; and (iii) scout bees are sent to investigate those locations.

In the research chapter that they published in 2015, Contreras-Cruz and his colleagues show how the ABC algorithm can be utilized to

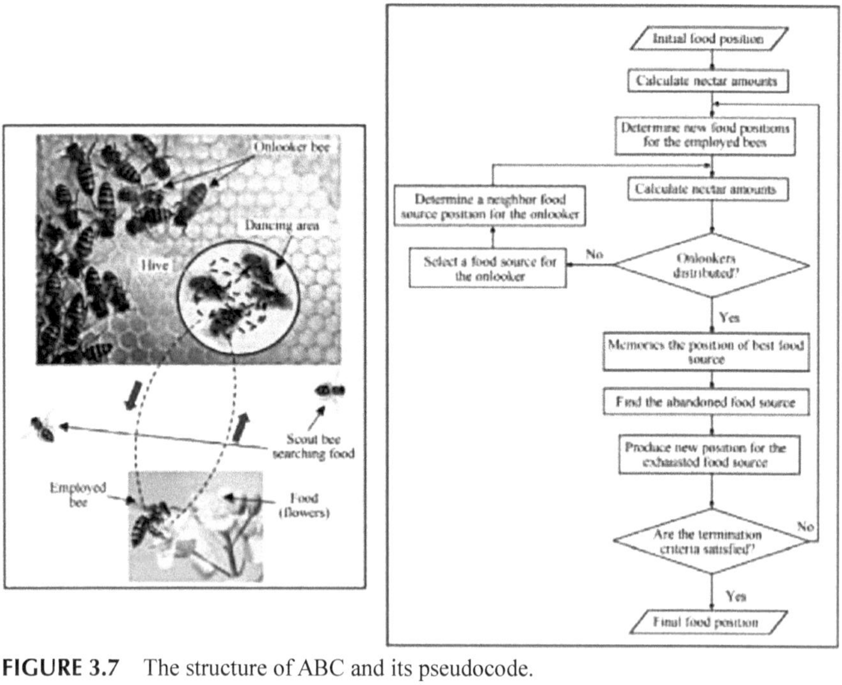

FIGURE 3.7 The structure of ABC and its pseudocode.

Source: Reprinted with permission from Ref.[13]

navigate mobile robots through predetermined environments. The method that was developed makes use of ABC for a local search and an evolutionary algorithm in order to determine the most effective way to proceed. For the purposes of verification, indoor experiments that take place in real time are provided. Despite the fact that the findings of Saffari & Mahjoob [68] can only be used in simulation settings, the authors describe a method that can also be used in a static environment. Ma & Lei [69] present the ABC-based technique as a means of navigating through a dynamic and real-time environment in order to get where one needs to go. They came up with a hybrid approach by combining the ABC algorithm with a time-rolling window method in order to come up with it. Despite the difficulty of navigating multiple mobile robots in an environment, Bhattacharjee et al. [70]; and Liang & Lee [71] were able to successfully implement ABC in a static environment. This was accomplished in both of their studies. The ABC method is evaluated in a manner that is comparable for both aerial and underwater navigation [72, 73] as well as routine tasks for autonomous vehicles [72–74, 95]. When plotting the most efficient route for a UCAV to take through the air, it is necessary to take into account the dangers and restrictions of the battleground in all three dimensions. Coelho et al. [61] implemented a modified version of the ABC algorithm, which is specifically designed to assist in UCAV navigation, in order to solve this problem. UCAV stands for unmanned combat aerial vehicle. The accuracy of exploration is improved by adding a balance-evolution strategy (BES) to the ABC algorithm. This leads to more successful exploration. This helps to achieve a balance between the two extremes of local exploitation and global exploration that can be described as a happy medium. When utilizing BES, the data obtained from convergence is utilized extensively throughout each cycle. Several examples of the ABC algorithm's application in the field of defense are provided by Wang et al. [30]. Within the scope of this investigation into an unmanned helicopter, a variety of challenging tasks, including data collection, accurate measuring, and border patrol, were put through their paces. They were able to determine the values of the two linear models' unobservable parameters with the assistance of experimental flight data by using a novel identification technique that was based on a chaotic operator and an ABC controller. This technique was developed by them.

3.3.11 CUCKOO SEARCH (CS) ALGORITHM

A metaheuristic algorithm known as the CS algorithm was initially presented for the first time by Yang & Deb in the year 2009 [75]. The algorithm takes its cue from the slothful behavior of certain cuckoos, which causes them to lay their eggs in the nests of other species. As a result, the algorithm lays its eggs in the nests of other species. Yang asserts that the method solves optimization issues by adhering to the three guidelines that are outlined in the following paragraphs:

- Cuckoos only use one nest for their egg, and they pick it at random.
- The best nests will be passed down through the generations because they produce the highest quality eggs.
- With a constant supply of nesting sites, a cuckoo's egg has a probability of discovery from the host bird of pa (0,1). The parent bird has two options: either abandon the egg or leave the nest to start a new one.

The navigation of mobile robots requires significant increases in both speed and processing power. Because of its improved convergence rate and increased efficiency, the CS algorithm is frequently utilized in the solution of engineering optimization problems. The high-level flowchart of the CS algorithm can be found in Figure 3.8. Due to the fact that the CS algorithm was only recently developed, it is only utilized in a limited number of academic papers that deal with the planning of routes for mobile robots. Mohanty & Parhi [76] have presented the algorithm that can be used to navigate a wheeled robot through a predetermined environment. Studies, both simulated and conducted in real time, have been conducted on a wheeled robot working in a complex environment in which certain aspects of the environment remain unknown. It is becoming more and more apparent that the two different sets of data agree very well as the deviation error between the experiments and the simulations continues to decrease. The CS-based approach shows promise as a form of navigational strategy when utilized in conjunction with other types of strategies. It has been shown that an evolutionary algorithm can be used to solve the problem of planning 3D aerial routes, and this has been demonstrated. Differential evolution is used to its full potential in order to optimize the process of selecting cuckoos for the improved CS model so that they can participate as agents in the search for the optimal path.

3.3.12 SHUFFLED FROG LEAPING ALGORITHM (SFLA)

For instance, if you look at Figure 3.8, you can see a representation of a proposed metaheuristic optimization method that was inspired by the feeding behaviors of frogs [77]. The SFLA is becoming increasingly common in engineering optimization, which is one of its applications. It has a higher success rate than any other metaheuristic algorithm, fewer parameters, faster convergence, and better search capacity when dealing with uncertainty compared to any other metaheuristic algorithm. Since it was first proposed by Mohanty and Parhi in 2015 as a means of achieving superior navigational outcomes in an uncertain environment, the combination of CS and ANFIS has since become standard practice in the field of engineering to achieve this goal. Wang et al. [78] propose a further hybrid path planning strategy for a 3D unknown environment by combining the differential evolution algorithm with CS. The goal of this strategy is to accelerate the rate at which the world converges on a single solution. The increased convergence rate of the aerial robot makes it easier to investigate the three-dimensional surrounding environment.

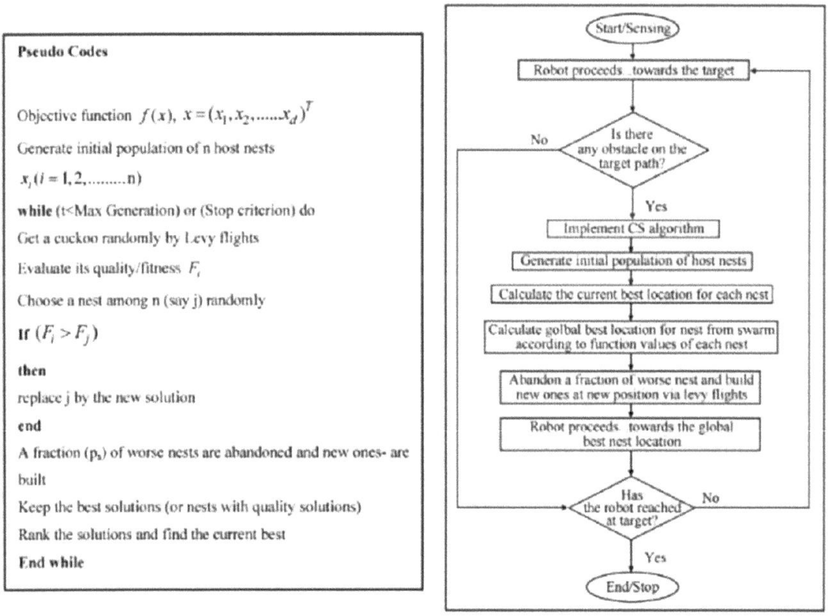

FIGURE 3.8 Pseudocode and flowchart of CS algorithm.

Source: Reprinted with permission from Ref.[76]

Gasparri & Prosperi [62] provided an illustration of a 3D environment that would be suitable for a battlefield and detailed how the CS algorithm could be implemented in such an environment. A prominent example of their work is the hybrid approach, which addresses issues related to both computer science and differential optimization and mobile robot navigation. Using SFLA, Hidalgo-Paniagua et al. [79] were able to develop a multi-objective technique for achieving navigational goals such as path safety, path length, and path smoothness. This technique was successful in achieving all of these objectives. Following this, we compare the newly developed method to the GA in a static environment and discover that the modified SFLA generates a much smoother path than the GA does. This chapter summarizes Parhi & Kundu's [80] research, in which they investigate and present a navigational strategy for use in three-dimensional underwater environments. For the purpose of tracking moving targets, they make use of an SFLA navigational strategy in their research, adapting it to the various conditions they find themselves in. By avoiding local minima, the adaptation method cuts down significantly on the amount of time and effort spent on navigation (Figure 3.9).

3.3.13 OTHER ALGORITHMS

Numerous intelligent techniques have been proposed by researchers to help mobile robots navigate in a wide range of environments, including the Invasive Weed Harmony Search (HS) Algorithm [81], the Optimization (IWO) [82], the Bat Algorithm (BA) [83], reactive obstacle avoidance based on boundary following using sliding mode control (Savkin & Wang).

3.4 CONCLUSION

After conducting an in-depth analysis of the research papers provided, navigational strategies were categorized as either classical or reactive. Only a few short decades ago, the vast majority of work in robotics still relied solely on traditional methods. Traditional methods have drawbacks such as high computational demands, the propensity to become mired in a "local minimum," a lack of scalability, an inability to deal with "maximum uncertainty," an inability to provide real-time navigation without extensive prior knowledge of the environment, and an inability to provide accurate sensory inputs in real time. In addition, traditional methods are unable to provide

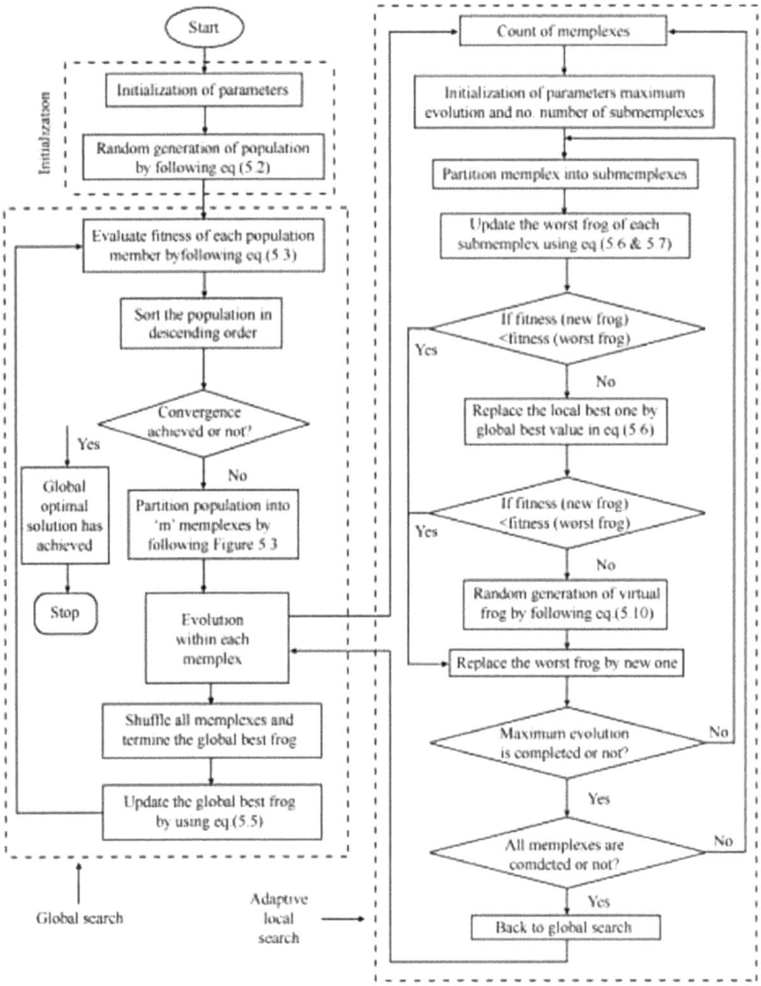

FIGURE 3.9 Flowchart of adaptive SFLA.

Source: Adapted from Ref.[80]

accurate sensory inputs in real time [13]. As a result of this, it is either unknown whether or not a solution can be attained through the use of the classical approach, or it is presumed that there is no such solution. Because of their inability to provide accurate predictions and consistent results, these methods are not robust enough to be utilized in a real-time setting. Although a great number of researchers have attempted to evaluate the drawbacks of classical techniques and developed new strategies, such as APF and hybrid algorithms, these methods still fall short of the capabilities offered by reactive approaches when it comes to practical applications. Classical techniques are typically utilized for navigation in environments that are already familiar to the user. This is due to the fact that classical techniques require prior knowledge of the working environment. Reactive strategies, on the other hand, are the ones that are used when traveling through uncharted territory because of their adaptability to different challenges. They are utilized for real-time navigational challenges as a result of their ease of implementation, ingenuity, and superior efficiency in comparison to conventional systems. However, reactive approaches still have some drawbacks that make them less desirable than classical ones. These drawbacks include a higher computing cost, a more complicated design, a learning phase, a need for a lot of memory, and the inability to be used with low-cost robots. Classical approaches do not have these drawbacks. Reactive methods are currently the method of choice because of their ability to rapidly adapt to an unknown setting while also requiring less computational effort than other methods. Recent developments in algorithm design, such as SFLA, PSO, ABC, and BFO, have proven to be effective for navigational tasks that involve merely repositioning obstacles. When used as a stand-alone controller, the real-time navigational path planning problem can be solved by any reactive or classical technique other than RA and CD. RA and CD are the only exceptions to this rule. To address the challenging issue of navigation, a number of researchers have proposed a hybrid navigational method. A review of the relevant literature reveals that all approaches, with the exception of SFLA, RA, and CD, effectively address the problem.

KEYWORDS

- **artificial intelligence**
- **autonomous system**
- **localization algorithms**
- **navigation**
- **positioning**
- **robot**
- **robotics**

REFERENCES

1. Borenstein, J., Everett, H. R., Feng, L., & Wehe, D., (1997). Mobile robot positioning: Sensors and techniques. *Journal of Robotic Systems, 14*(4). https://doi.org/10.1002/(SICI)1097-4563(199704)14:4<231::AID-ROB2>3.0.CO;2-R.
2. Siegwart, R., & Nourbakhsh, I., (2011). Mobile robot localization. *Introduction to Autonomous Mobile Robots.* MIT Press.
3. Chenavier, F., & Crowley, J. L., (1992). Position estimation for a mobile robot using vision and odometry. *Proceedings – IEEE International Conference on Robotics and Automation, 3.* https://doi.org/10.1109/robot.1992.220052.
4. Cox, I. J., (1991). Blanche – An experiment in guidance and navigation of an autonomous robot vehicle. *IEEE Transactions on Robotics and Automation, 7*(2). https://doi.org/10.1109/70.75902.
5. Borenstein, J., Everett, H. R., Feng, L., & Wehe, D. (1997). Mobile robot positioning: Sensors and techniques. *Journal of Robotic Systems, 14*(4), 231–249.
6. Esteves, J. S., Carvalho, A., & Couto, C., (2003). Generalized geometric triangulation algorithm for mobile robot absolute self-localization. *IEEE International Symposium on Industrial Electronics, I.* https://doi.org/10.1109/ISIE.2003.1267272.
7. Milos, S., (2007). *Roadmap Methods vs. Cell Decomposition in Robot Motion Planning*, 127–132.
8. Seda, M., & Pich, V., (2008). Robot motion planning using generalized Voronoi diagrams. *ISCGAV'08: Proceedings of the 8th WSEAS International Conference on Signal Processing, Computational Geometry and Artificial Vision.*
9. Schwartz, J. T., & Sharir, M., (1983). On the "piano movers" problem I. The case of a two-dimensional rigid polygonal body moving amidst polygonal barriers. *Communications on Pure and Applied Mathematics, 36*(3), 345–398. https://doi.org/https://doi.org/10.1002/cpa.3160360305.
10. Choset, H., & Burdick, J., (2000). Sensor-based exploration: The hierarchical generalized Voronoi graph. *International Journal of Robotics Research, 19*(2). https://doi.org/10.1177/02783640022066770.
11. De Berg, M., Cheong, O., Van, K. M., & Overmars, M., (2008). *Computational Geometry: Algorithms and Applications*. https://doi.org/10.1007/978-3-540-77974-2.
12. Howie, C., & Kevin, L. S. H., (2005). Principles of robot motion. In: *The Canadian Nurse* (Vol. 53).
13. Patle, B. K., Babu, L. G., Pandey, A., Parhi, D. R. K., & Jagadeesh, A., (2019). A review: On path planning strategies for navigation of mobile robot. In: *Defense Technology* (Vol. 15, No. 4). https://doi.org/10.1016/j.dt.2019.04.011.
14. Khatib, O., (1985). Real-time obstacle avoidance for manipulators and mobile robots. *Proceedings – IEEE International Conference on Robotics and Automation.* https://doi.org/10.1109/ROBOT.1985.1087247.
15. Garibotto, G., & Masciangelo, S., (2002). *Path Planning Using the Potential Field Approach for Navigation.* https://doi.org/10.1109/icar.1991.240366.
16. Kim, J. O., & Khosla, P. K., (1992). Real-time obstacle avoidance using harmonic potential functions. *IEEE Transactions on Robotics and Automation, 8*(3). https://doi.org/10.1109/70.143352.

17. Borenstein, J., & Koren, Y., (1989). Real-time obstacle avoidance for fast mobile robots. *IEEE Transactions on Systems, Man and Cybernetics, 19*(5). https://doi.org/10.1109/21.44033.
18. Ge, S. S., & Cui, Y. J., (2002). Dynamic motion planning for mobile robots using potential field method. *Autonomous Robots, 13*(3). https://doi.org/10.1023/A:1020564024509.
19. Valavanis, K. P., Hebert, T., Kolluru, R., & Tsourveloudis, N., (2000). Mobile robot navigation in 2-D dynamic environments using an electrostatic potential field. *Systems, Man and Cybernetics, Part A: Systems and Humans, IEEE Transactions, 30*, 187–196. https://doi.org/10.1109/3468.833100.
20. Huang, L., (2009). Velocity planning for a mobile robot to track a moving target – a potential field approach. *Robotics and Autonomous Systems, 57*(1). https://doi.org/10.1016/j.robot.2008.02.005.
21. Shi, P., & Zhao, Y., (2009). An efficient path planning algorithm for mobile robot using improved potential field. In: *2009 IEEE International Conference on Robotics and Biomimetics; ROBIO 2009*. https://doi.org/10.1109/ROBIO.2009.5420407.
22. Sfeir, J., Saad, M., & Saliah-Hassane, H., (2011). An improved artificial potential field approach to real-time mobile robot path planning in an unknown environment. *ROSE 2011 – IEEE International Symposium on Robotic and Sensors Environments, Proceedings*. https://doi.org/10.1109/ROSE.2011.6058518.
23. Bremermann, H. J. (1958). *The Evolution Of Intelligence: The Nervous System as a Model of Its Environment*. University of Washington, Department of Mathematics.
24. Holland, J. H., (2019). *Adaptation in Natural and Artificial Systems*. https://doi.org/10.7551/mitpress/1090.001.0001.
25. Shing, M., & Parker, G. B., (1993). *Genetic Algorithms for the Development of Real-Time Multi-Heuristic Search Strategies*. ICGA.
26. Kang, X., Yue, Y., Li, D., & Maple, C., (2011). Genetic algorithm-based solution to dead-end problems in robot navigation. *International Journal of Computer Applications in Technology, 41*(3, 4). https://doi.org/10.1504/IJCAT.2011.042693.
27. Shi, P., & Cui, Y., (2010). Dynamic path planning for mobile robot based on genetic algorithm in unknown environment. In: *2010 Chinese Control and Decision Conference; CCDC 2010*. https://doi.org/10.1109/CCDC.2010.5498349.
28. Pratihar, D. K., Deb, K., & Ghosh, A., (1999). Fuzzy-genetic algorithms and time-optimal obstacle-free path generation for mobile robots. *Engineering Optimization, 32*(1). https://doi.org/10.1080/03052159908941294.
29. Hui, N. B., & Pratihar, D. K., (2009). A comparative study on some navigation schemes of a real robot tackling moving obstacles. *Robotics and Computer-Integrated Manufacturing, 25*(4, 5). https://doi.org/10.1016/j.rcim.2008.12.003.
30. Wang, X., Shi, Y., Ding, D., & Gu, X., (2016). Double global optimum genetic algorithm-particle swarm optimization-based welding robot path planning. *Engineering Optimization, 48*(2). https://doi.org/10.1080/0305215X.2015.1005084.
31. Kala, R., (2014). Coordination in navigation of multiple mobile robots. *Cybernetics and Systems, 45*(1). https://doi.org/10.1080/01969722.2014.862085.
32. Liu, F., Liang, S., & Xian, X., (2014). Optimal robot path planning for multiple goals visiting based on tailored genetic algorithm. *International Journal of Computational Intelligence Systems, 7*(6). https://doi.org/10.1080/18756891.2014.963978.

33. Zadeh, L. A., (1965). Fuzzy sets-information and control-1965. *Information and Control, 8.*
34. Zavlangas, P. G., & Tzafestas, S. G., (2003). Motion control for mobile robot obstacle avoidance and navigation: A fuzzy logic-based approach. *Systems Analysis Modelling Simulation, 43*(12). https://doi.org/10.1080/0232929032000115100.
35. Tschichold-Gürman, N., (1997). The neural network model RuleNet and its application to mobile robot navigation. *Fuzzy Sets and Systems, 85*(2), 287–303. https://doi.org/https://doi.org/10.1016/0165-0114(95)00351-7.
36. Homaifar, A., & McCormick, E., (1995). Simultaneous design of membership functions and rule sets for fuzzy controllers using genetic algorithms. *IEEE Transactions on Fuzzy Systems, 3*(2). https://doi.org/10.1109/91.388168.
37. Jaradat, M. A. K., Garibeh, M. H., & Feilat, E. A., (2012). Autonomous mobile robot dynamic motion planning using hybrid fuzzy potential field. *Soft Computing, 16*(1). https://doi.org/10.1007/s00500-011-0742-z.
38. Yen, C. T., & Cheng, M. F., (2018). A study of fuzzy control with ant colony algorithm used in mobile robot for shortest path planning and obstacle avoidance. *Microsystem Technologies, 24*(1). https://doi.org/10.1007/s00542-016-3192-9.
39. Al-Khatib, M., & Saade, J. J., (2003). An efficient data-driven fuzzy approach to the motion planning problem of a mobile robot. *Fuzzy Sets and Systems, 134*(1). https://doi.org/10.1016/S0165-0114(02)00230-0.
40. Lee, H., Jung, J., Choi, K., Park, J., & Myung, H., (2012). Fuzzy-logic-assisted interacting multiple model (FLAIMM) for mobile robot localization. *Robotics and Autonomous Systems, 60*(12). https://doi.org/10.1016/j.robot.2012.09.018.
41. Hoy, M., Matveev, A. S., & Savkin, A. V., (2012). Collision free cooperative navigation of multiple wheeled robots in unknown cluttered environments. *Robotics and Autonomous Systems, 60*(10). https://doi.org/10.1016/j.robot.2012.07.002.
42. Kang, T. K., Zhang, H., Park, G. T., & Kim, D. W., (2013). Ego-motion-compensated object recognition using type-2 fuzzy set for a moving robot. *Neurocomputing, 120.* https://doi.org/10.1016/j.neucom.2012.09.041.
43. Al-Mutib, K., Mattar, E., & Alsulaiman, M., (2015). Implementation of fuzzy decision based mobile robot navigation using stereo vision. *Procedia Computer Science, 62.* https://doi.org/10.1016/j.procs.2015.08.427.
44. Nazari, M. A. D., & Khooban, M. H., (2015). Design of optimal Mamdani-type fuzzy controller for nonholonomic wheeled mobile robots. *Journal of King Saud University – Engineering Sciences, 27*(1). https://doi.org/10.1016/j.jksues.2013.05.003.
45. Castillo, O., Neyoy, H., Soria, J., Melin, P., & Valdez, F., (2015). A new approach for dynamic fuzzy logic parameter tuning in ant colony optimization and its application in fuzzy control of a mobile robot. *Applied Soft Computing Journal, 28.* https://doi.org/10.1016/j.asoc.2014.12.002.
46. Qiao, J., Fan, R., Han, H., & Ruan, X., (2009). Q-learning based on dynamical structure neural network for robot navigation in unknown environment. *Lecture Notes in Computer Science (Including Subseries Lecture Notes in Artificial Intelligence and Lecture Notes in Bioinformatics), 5553 LNCS (PART 3).* https://doi.org/10.1007/978-3-642-01513-7_21.

47. Li, Q. L., Song, Y., & Hou, Z. G., (2015). Neural network based FastSLAM for autonomous robots in unknown environments. *Neurocomput., 165*(C), 99–110. https://doi.org/10.1016/j.neucom.2014.06.095.
48. Na, Y. K., & Oh, S. Y., (2003). Hybrid control for autonomous mobile robot navigation using neural network based behavior modules and environment classification. *Autonomous Robots, 15*(2). https://doi.org/10.1023/A:1025597227189.
49. Eberhart, R., & Kennedy, J., (1995). New optimizer using particle swarm theory. *Proceedings of the International Symposium on Micro Machine and Human Science.* https://doi.org/10.1109/mhs.1995.494215.
50. Tang, X. L., Li, L. M., & Jiang, B. J., (2014). Mobile robot SLAM method based on multi-agent particle swarm optimized particle filter. *Journal of China Universities of Posts and Telecommunications, 21*(6). https://doi.org/10.1016/S1005-8885(14)60348-4.
51. Ha, X. V., Ha, C., & Lee, J., (2013). Novel hybrid optimization algorithm using PSO and MADS for the trajectory estimation of a four-track wheel skid-steered mobile robot. *Advanced Robotics, 27*(18). https://doi.org/10.1080/01691864.2013.839089.
52. Atyabi, A., Phon-Amnuaisuk, S., & Ho, C. K., (2010). Applying area extension PSO in robotic swarm. *Journal of Intelligent and Robotic Systems: Theory and Applications, 58*(3, 4). https://doi.org/10.1007/s10846-009-9374-2.
53. Tang, Q., & Eberhard, P., (2011). Cooperative motion of swarm mobile robots based on particle swarm optimization and multibody system dynamics. *Mechanics Based Design of Structures and Machines, 39*(2). https://doi.org/10.1080/15397734.2011.550855.
54. Couceiro, M. S., Rocha, R. P., & Ferreira, N. M. F., (2013). A PSO multi-robot exploration approach over unreliable Manets. *Advanced Robotics, 27*(16). https://doi.org/10.1080/01691864.2013.819605.
55. Dorigo, M., & Gambardella, L. M., (1997). Ant colony system: A cooperative learning approach to the traveling salesman problem. *IEEE Transactions on Evolutionary Computation, 1*(1). https://doi.org/10.1109/4235.585892.
56. Tan, G. Z., He, H., & Sloman, A., (2007). Ant colony system algorithm for real-time globally optimal path planning of mobile robots. *Zidonghua Xuebao/Acta Automatica Sinica, 33*(3). https://doi.org/10.1360/aas-007-0279.
57. Liu, J., Yang, J., Liu, H., Tian, X., & Gao, M., (2017). An improved ant colony algorithm for robot path planning. *Soft Computing, 21*(19). https://doi.org/10.1007/s00500-016-2161-7.
58. Kumar, P. B., Sahu, C., & Parhi, D. R., (2018). A hybridized regression-adaptive ant colony optimization approach for navigation of humanoids in a cluttered environment. *Applied Soft Computing Journal, 68.* https://doi.org/10.1016/j.asoc.2018.04.023.
59. Passino, K. M., (2002). Biomimicry of bacterial foraging for distributed optimization and control. *IEEE Control Systems, 22*(3). https://doi.org/10.1109/MCS.2002.1004010.
60. dos Santos Coelho, L., & Sierakowski, C. A. (2005). Bacteria Colony Approaches With Variable Velocity Applied To Path Optimization of Mobile Robots.
61. Sierakowski, C. A., & Coelho, L. D. S., (2006). Path planning optimization for mobile robots based on bacteria colony approach. *Advances in Soft Computing, 34.* https://doi.org/10.1007/3-540-31662-0_15.

62. Gasparri, A., & Prosperi, M., (2008). A bacterial colony growth algorithm for mobile robot localization. *Autonomous Robots*, *24*(4). https://doi.org/10.1007/s10514-007-9076-1.
63. Abbas, N. H., & Ali, F. M., (2017). Path planning of an autonomous mobile robot using enhanced bacterial foraging optimization algorithm. *Al-Khwarizmi Engineering Journal*, *12*(4). https://doi.org/10.22153/kej.2016.01.001.
64. Jati, A., Singh, G., Rakshit, P., Konar, A., Kim, E., & Nagar, A. K., (2012). A hybridization of improved harmony search and bacterial foraging for multi-robot motion planning. *2012 IEEE Congress on Evolutionary Computation, CEC 2012*. https://doi.org/10.1109/CEC.2012.6256533.
65. Oyekan, J., & Hu, H., (2010). A novel bacterial foraging algorithm for automated tuning of PID controllers of UAVs. In: *2010 IEEE International Conference on Information and Automation, ICIA 2010*. https://doi.org/10.1109/ICINFA.2010.5512477.
66. Tan, L., Wang, H., Yang, C., & Niu, B., (2017). A multi-objective optimization method based on discrete bacterial algorithm for environmental/economic power dispatch. *Natural Computing*, *16*(4). https://doi.org/10.1007/s11047-017-9620-7.
67. Karaboga, D., (2005). *An Idea Based on Honey Bee Swarm for Numerical Optimization*. Technical Report TR06, Erciyes University, TR06.
68. Saffari, M. H., & Mahjoob, M. J., (2009). Bee colony algorithm for real-time optimal path planning of mobile robots. *ICSCCW 2009 – 5th International Conference on Soft Computing, Computing with Words and Perceptions in System Analysis, Decision and Control*. https://doi.org/10.1109/ICSCCW.2009.5379462.
69. Ma, Q., & Lei, X., (2010). Dynamic path planning of mobile robots based on ABC algorithm. *Lecture Notes in Computer Science (Including Subseries Lecture Notes in Artificial Intelligence and Lecture Notes in Bioinformatics)*, *6320 LNAI* (PART 2). https://doi.org/10.1007/978-3-642-16527-6_34.
70. Bhattacharjee, P., Rakshit, P., Goswami, I., Konar, A., & Nagar, A. K., (2011). Multi-robot path-planning using artificial bee colony optimization algorithm. *Proceedings of the 2011 3rd World Congress on Nature and Biologically Inspired Computing, NaBIC 2011*. https://doi.org/10.1109/NaBIC.2011.6089601.
71. Liang, J. H., & Lee, C. H., (2015). Efficient collision-free path-planning of multiple mobile robots system using efficient artificial bee colony algorithm. *Advances in Engineering Software*, *79*. https://doi.org/10.1016/j.advengsoft.2014.09.006.
72. Xu, C., Duan, H., & Liu, F., (2010). Chaotic artificial bee colony approach to uninhabited combat air vehicle (UCAV) path planning. *Aerospace Science and Technology*, *14*(8). https://doi.org/10.1016/j.ast.2010.04.008.
73. Li, B., Chiong, R., & Gong, L. G., (2014). Search-evasion path planning for submarines using the artificial bee colony algorithm. *Proceedings of the 2014 IEEE Congress on Evolutionary Computation, CEC 2014*. https://doi.org/10.1109/CEC.2014.6900224.
74. Bhagade, A. S., & Puranik, P. V., (2012). Artificial bee colony (ABC) algorithm for vehicle routing optimization problem. *International Journal of Soft Computing and Engineering (IJSCE)*, *2*(2).
75. Yang, X. S., & Deb, S., (2009). Cuckoo search via Lévy flights. In: *2009 World Congress on Nature and Biologically Inspired Computing, NABIC 2009 – Proceedings*. https://doi.org/10.1109/NABIC.2009.5393690.

76. Mohanty, P. K., & Parhi, D. R., (2016). Optimal path planning for a mobile robot using cuckoo search algorithm. *Journal of Experimental and Theoretical Artificial Intelligence*, *28*(1, 2). https://doi.org/10.1080/0952813X.2014.971442.
77. Eusuff, M. M., & Lansey, K. E., (2003). Optimization of water distribution network design using the shuffled frog leaping algorithm. *Journal of Water Resources Planning and Management*, *129*(3). https://doi.org/10.1061/(asce)0733-9496(2003)129:3(210).
78. Wang, G., Guo, L., Duan, H., Wang, H., Liu, L., & Shao, M., (2012). A hybrid metaheuristic DE/CS algorithm for UCAV three-dimension path planning. *The Scientific World Journal*, *2012*. https://doi.org/10.1100/2012/583973.
79. Hidalgo-Paniagua, A., Vega-Rodríguez, M. A., Ferruz, J., & Pavón, N., (2015). MOSFLA-MRPP: Multi-objective shuffled frog-leaping algorithm applied to mobile robot path planning. *Engineering Applications of Artificial Intelligence*, *44*. https://doi.org/10.1016/j.engappai.2015.05.011.
80. Parhi, D. R., & Kundu, S., (2017). Navigational strategy for underwater mobile robot based on adaptive neuro-fuzzy inference system model embedded with shuffled frog leaping algorithm–based hybrid learning approach. *Proceedings of the Institution of Mechanical Engineers Part M: Journal of Engineering for the Maritime Environment*, *231*(4). https://doi.org/10.1177/1475090216684235.
81. Mohanty, P. K., & Parhi, D. R., (2014). A new efficient optimal path planner for mobile robot based on invasive weed optimization algorithm. *Frontiers of Mechanical Engineering*, *9*(4). https://doi.org/10.1007/s11465-014-0304-z.
82. Kundu, S., & Parhi, D. R., (2016). Navigation of underwater robot based on dynamically adaptive harmony search algorithm. *Memetic Computing*, *8*(2). https://doi.org/10.1007/s12293-016-0179-0.
83. Guo, J., Gao, Y., & Cui, G., (2015). The path planning for mobile robot based on bat algorithm. *International Journal of Automation and Control*, *9*(1). https://doi.org/10.1504/IJAAC.2015.068041.
84. Castellano, G., Attolico, G., & Distante, A., (1997). Automatic generation of fuzzy rules for reactive robot controllers. *Robotics and Autonomous Systems*, *22*(2). https://doi.org/10.1016/S0921-8890(97)00036-5.
85. Chakraborty, J., Konar, A., Chakraborty, U. K., & Jain, L. C., (2008). Distributed cooperative multi-robot path planning using differential evolution. In: *2008 IEEE Congress on Evolutionary Computation, CEC 2008*. https://doi.org/10.1109/CEC.2008.4630875.
86. Contreras-Cruz, M. A., Ayala-Ramirez, V., & Hernandez-Belmonte, U. H., (2015). Mobile robot path planning using artificial bee colony and evolutionary programming. *Applied Soft Computing*, *30*. https://doi.org/10.1016/j.asoc.2015.01.067.
87. Janglová, D., (2004). Neural networks in mobile robot motion. *International Journal of Advanced Robotic Systems*, *1*(1). https://doi.org/10.5772/5615.
88. Kareem, J. M. A., Al-Rousan, M., & Quadan, L., (2011). Reinforcement based mobile robot navigation in dynamic environment. *Robotics and Computer-Integrated Manufacturing*, *27*(1). https://doi.org/10.1016/j.rcim.2010.06.019.
89. Mohanty, P. K., & Parhi, D. R., (2015). A new hybrid optimization algorithm for multiple mobile robots navigation based on the CS-ANFIS approach. *Memetic Computing*, *7*(4). https://doi.org/10.1007/s12293-015-0160-3.

90. Savkin, A. V., & Wang, C., (2013). A simple biologically inspired algorithm for collision-free navigation of a unicycle-like robot in dynamic environments with moving obstacles. *Robotica, 31*(6). https://doi.org/10.1017/S0263574713000313.
91. Thanh, H. L. N. N., Phi, N. N., & Hong, S. K., (2018). Simple nonlinear control of quadcopter for collision avoidance based on geometric approach in static environment. *International Journal of Advanced Robotic Systems, 15*(2). https://doi.org/10.1177/1729881418767575.
92. Kang, T. K., Zhang, H., Park, G. T., & Kim, D. W. (2013). Ego-motion-compensated object recognition using type-2 fuzzy set for a moving robot. *Neurocomputing, 120*, 130–140.
93. Chen, Y. L., Cheng, J., Lin, C., Wu, X., Ou, Y., & Xu, Y. (2013). Classification-based learning by particle swarm optimization for wall-following robot navigation. *Neurocomputing, 113*, 27–35.
94. Dorigo, M., & Gambardella, L. M. (1997). Ant colony system: a cooperative learning approach to the traveling salesman problem. *IEEE Transactions on evolutionary computation, 1*(1), 53–66.
95. Xu, C., Duan, H., & Liu, F. (2010). Chaotic artificial bee colony approach to Uninhabited Combat Air Vehicle (UCAV) path planning. *Aerospace science and technology, 14*(8), 535–541.

CHAPTER 4

Methodologies for Controlling the Autonomous System

HITESH KUMAR SHARMA, RAHUL GUSAIN, and DEVYANSH SAMYAL

Cybernetics Cluster, School of Computer Science and Engineering, University of Petroleum and Energy Studies, Dehradun, Uttarakhand, India

ABSTRACT

Artificial Intelligence, Internet of Things, Image Processing, Computer Vision, Cloud Computing Network, Data Science, and DevOps are some advanced IT technologies which have made autonomous systems a reality. Controlling an autonomous system requires a series of steps to be performed in a sequential manner. An efficient methodology to control an autonomous system is necessary for successful implementation in the real world. Auto-Controlled systems are required in many advanced fields like smart cities, smart agriculture, smart healthcare, smart transport, etc. The efficient functioning of such auto-controlled systems is equally important. In this chapter, we have described efficient methodologies for Autonomous Systems. The challenges faced by autonomous systems and their solutions will also be discussed in this chapter.

4.1 INTRODUCTION

A network or group of networks that are all administered and overseen by a single body or organization is referred to as an autonomous system (AS).

Advances in Autonomous Navigation Through Intelligent Technologies.
Niharika Singh, Thipendra P. Singh, Brian Azzopardi (Eds.)

An AS is a diverse network that is often under the control of a significant company. There are several separate subnetworks inside an AS that share common routing regulations and unified routing algorithms. The autonomous system may be used with class A, class B, or class C networks and is organized around IP prefixes [1]. When used with AS, the word "prefix" refers to a "Classless Inter-Domain Routing (CIDR) block," which is a collection of IP addresses with the same prefix and the same amount of bits. Because of this, a routing domain is another name for an autonomous system. Each AS is in charge of the IP address space determined by its IP prefixes and exchanges routing information depending on its routing regulations. In this manner, a device in a network in one AS and a device in a network in another AS can exchange data packets [2]. Automation and autonomy complement one another, strengthening and reverting to one another as necessary. Consider a remotely controlled ferry that is driven from the beach, for instance (this is automation as opposed to autonomy). An autonomous system that will get the ferry to a safe state is required as a backup in the event that connection with the shore is lost for whatever reason. Only until low-cost sensors, the ability to manage massive volumes of data, and the computing power and tools to execute the essential decision algorithms are available will real autonomously control systems be practical. "Mixed" autonomous control systems come with significant dangers (Figure 4.1).

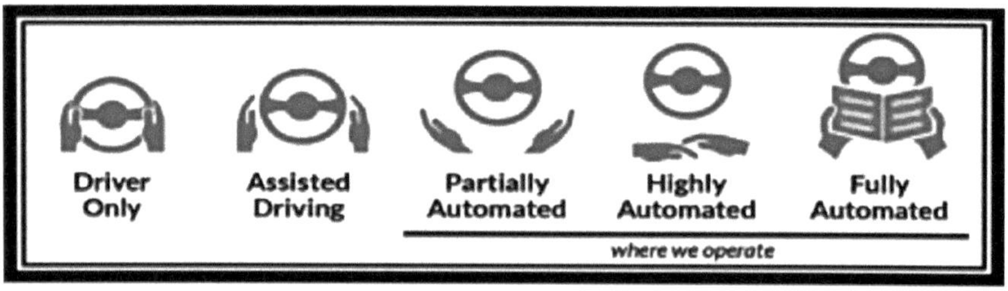

FIGURE 4.1 Level of autonomous system.

For instance, autonomous driving features may be advertised as a safety aid, but once users are made aware of them, their driving patterns may alter, and they may place more reliance than is appropriate in these safety features. This is truly dangerous since it increases the total risk for a driver on the road when there isn't complete autonomy. This difficulty ought to be resolved after complete autonomy (Level 5) has been attained.

4.2 AUTONOMOUS SYSTEM

Autonomous systems are huge networks that form the internet. An autonomous system is a big group or a collection of networks that follow similar routing protocols. Any device that has a connection to the internet makes use of an autonomous system. It is assigned a 16-digit number that is unique and functions globally; it is mainly used for identification. It is popularly known as the autonomous system number or ASN [3]. The assignment of this number is done by the Internet Assigned Numbers Authority (IANA). It is estimated that there are about 90,000 ASNs being used globally. IANA administers the distribution of ASNs across various global regions. Each of these regions controls a self-regional internet registry, which in turn is used for issuing an ASN for an individual autonomous system. For a better understanding, we can consider an example of a post office; an Autonomous system will be like our town's post office, where any mail posted goes from one post office to another until it reaches the correct town, and then the town's post office delivers the mail to the correct destination within the town itself. Similarly, data packets travel by hopping from one autonomous system to another until they reach the autonomous system that contains their destination IP address. The routers that are available within that autonomous system send the packets to the IP address. Every autonomous system has control over a certain set of IP addresses, and this specific range of IP addresses is known as their "IP address space." Autonomous system numbers are of great significance as they tell us what ASN is used for what purpose. IANA has made autonomous systems starting from 1 to 64,511 available for use globally, whereas the series beginning from 64,512 to 65,535 is reserved for any private purpose [4]. The introduction of the autonomous system was done to administer the organizations associated with networking like ISPs (internet service providers), government-owned agencies, and educational institutions. An autonomous system is also described as a group of routers administered by a single entity. Networks inside an autonomous system communicate with each other using gateway protocols. They use the Border Gateway Protocol, which manages the routing of packets among different autonomous systems across the internet. BGP uses the autonomous system number (ASN) for identification.

4.3 APPLICATION OF AUTONOMOUS SYSTEM

4.3.1 ADVANCED SENSORS

Modern sensors are an essential part of autonomous systems. They help in precise observation of the surroundings and gather data through these observations. This data is further analyzed, and necessary plans are carried out. For example, the sensors installed in an autonomous vehicle will help in determining distances and other information about obstacles like other vehicles, pedestrians, etc. This will minimize the chances of accidents. Some of the common types of sensors used in autonomous systems are camera sensors, ultrasonic sensors, infrared sensors, radar sensors, gyro sensors, IMU sensors, etc. [5, 6]. Usually, in an autonomous system, a combination of sensors, known as sensor fusion, is used to provide more accurate results. Sensor fusion helps in achieving improved data management, machine learning, decision making, etc.

4.3.2 WIRELESS COMMUNICATION

Rapid and easy communication is a major aspect of autonomous systems. Well-developed machines with improved networking and sensing provide effortless communication to transfer information [7]. For example, in an autonomous vehicle, it can communicate with the devices present in the vehicle, other nearby vehicles to avoid accidents, some network to transfer information, pedestrians via smartphones, parking systems, or traffic signals, etc.

4.3.3 AUTONOMOUS ROBOTS

Various types of autonomous robots are being developed around the world. Basic robots like vacuum cleaning machines are commonly used in most households. Whereas giant autonomous robots are used for lifting heavy objects in construction industries. Autonomous robots are also helpful in medical surgeries and operations for precise and careful treatment procedures. Autonomous vehicles come with great features and functionalities like artificial intelligence, sensors, high security, etc. UAVs [7, 8] and drones are one of the most popular autonomous machines. They mostly work on an auto-pilot mode. Sensors and monitors present in drones make them a magnificent tool for exploring, researching, and studying environments.

4.3.4 SAFE SYSTEMS

Safety standards are considered with care while building autonomous systems; otherwise, it would be unsafe for the customers and the concerned industries as well. Autonomous systems mostly show a better result in terms of safety and cover for man-made mistakes, which makes them reliable. For example, autonomous vehicles have seat belt warnings, wrong lane warnings, obstacle sensors, and many other features to avoid danger. Some software that is used to make autonomous systems safe are Hypervisor, Secure data communications and over-the-air updates (OTA), RTOS, Advanced driver assistance systems (ADAS) [9], etc.

4.4 TECHNOLOGIES INVOLVED IN CONTROLLING AN AUTONOMOUS SYSTEM

Sensors and sensor fusion are crucial in autonomous systems. They make it possible for these systems to collect information from the environment's sources and use the information at hand to organize and carry out an activity. This section discusses the many sensor types used by autonomous systems. Sensors enable an autonomous system to locate itself, identify objects, gather information, and avoid dangers [10, 11]. The common types of sensors and their functions are described below, along with an explanation of how using numerous sensors can make up for a particular type of sensor's shortcoming by combining it with a different type of sensor that has various strengths and weaknesses (Figure 4.2).

4.4.1 CAMERA SENSORS

Common optical sensor types include cameras that capture still images and video. In order to recognize vehicles, pedestrians, traffic signs and signals, guardrails, and road markings in a traffic situation, many cameras can offer a 360° view of the surroundings. Camera sensors are as important for autonomy in other settings, such as shops and warehouses.

4.4.2 ULTRASONIC SENSORS

High-frequency sound waves are used by ultrasonic sensors to calculate the distance between the sensor and an item. Short-range applications, like a car's parking assistance system, frequently use ultrasonic sensors.

4.4.3 INFRARED SENSORS

Night-vision systems use infrared sensors to produce images in low-light situations.

4.4.4 LIDAR SENSORS

LiDAR sensors use lasers to measure distance and build highly detailed 3D models of their surroundings. For mass-market application, rotating LiDAR systems that provide a complete 360° view around an autonomous device are too expensive. However, some less expensive solid-state LiDAR systems [12] with a constrained field of view are practical for widespread deployment. Radar detectors for better distance control and braking aid applications, radar sensors in autonomous vehicles detect blind spots using short-range (24 GHz) and long-range (77 GHz) radio waves [13].

FIGURE 4.2 Functions of autonomous system.

4.5 CHALLENGES IN CONTROLLING AN AUTONOMOUS SYSTEM

Autonomous systems are capable of processing the given information on their own. These systems can make decisions independently and execute them. Research on this subject has not focused on integrating planning systems, but rather on the effects of autonomous control methods. Therefore, with the current information available, it is not possible to provide substantial data to recommend a combination of technologies that can achieve a sufficient level of logistical success in production systems of varying complexity. A three-dimensional evaluation system is expected to control autonomous systems. It includes the following subsections.

4.5.1 COMPLEXITY

Wind et al. present the most widely used approach to complexity. The methods for controlling a system are partly determined by its complexity [14]. Although complexity is included in the author's rating system, it remains at a relative level, making direct comparisons between different systems difficult. Figure 4.3 explains the various categories of complexity.

4.5.2 AUTONOMY

Böse & Windt introduced the first method for governing autonomy in logistic systems. To "enable a first approximation examination of autonomously controlled logistic order processing," the authors list criteria. Factors such as the extent, scope, and level of integration between planning and control systems play a role. The feedback loop between planning and control methods indicates how quickly information is exchanged and processed. The accompanying graphic lists various factors and autonomous patterns.

4.5.3 LOGISTIC OBJECTIVE ACHIEVEMENT

Due date reliability, throughput time, utilization, and work in progress are the four factors recommended by Wind et al. for measuring the logistic objective achievement of autonomous control methods. The achievement of each logistics aim can be calculated because each objective is weighted. The end result is a percentage figure that represents the fulfillment of the logistical objective [15]. The definition of production scenarios with varying levels of complexity is made possible by the complexity evaluation pattern. Different coupling strategies can be specified and ranked according to their degree of autonomy using the autonomy assessment pattern. As a result, each simulation research can be placed in a specific location within the complexity and autonomy spectrum. Finally, the construction of a demonstrable curve will be possible thanks to the measurement of logistic objective achievement (Figure 4.3).

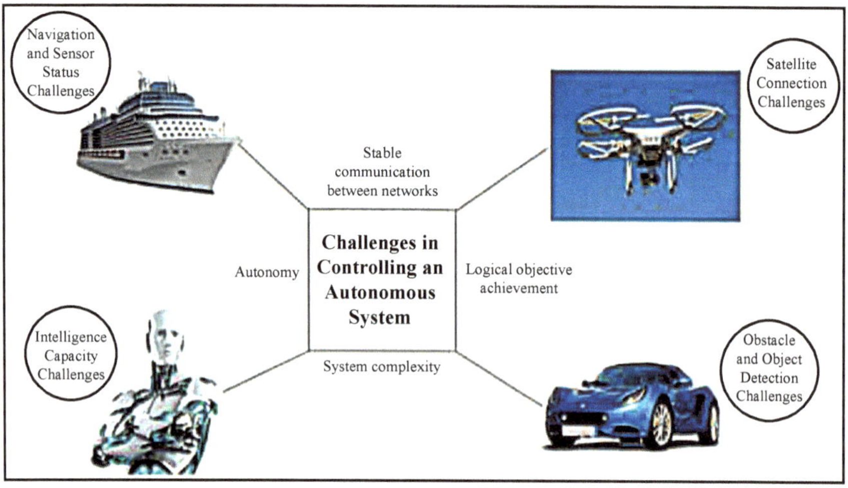

FIGURE 4.3 Challenges in controlling autonomous system.

4.6 FUTURE APPLICATION OF AUTONOMOUS SYSTEM

The number of parameters rises, and the logic, causes, and consequences of decision-making get more complex once autonomous systems are making judgments and taking actions. It's critical to comprehend how to guarantee the objectivity, security, accountability, and morality of machine-based decision-making. Failures will happen no matter how skillfully we design and develop our systems. That may imply that an autonomous car hits a stationary object, causes an accident, or results in fatalities. Safety and legal responsibilities go hand in hand. Who is in charge if the car wrecks?

Which is to blame: the human operator, the automaker, the programmer, the AI, or the network connection? Who assumes the insurance risk when the judgments made by machines affect human life? Before widespread adoption can take place, these legal obligations must be resolved. This has to do with the morals of AI. How can an autonomous system decide what is "best" or "correct" if it uses learning algorithms to secure the best results? It is important to think about issues like sharing responsibility, ethical use of technology and data, and AI [16]. To ensure that various systems can interact, communicate, and contribute value, standards are required (Figure 4.4).

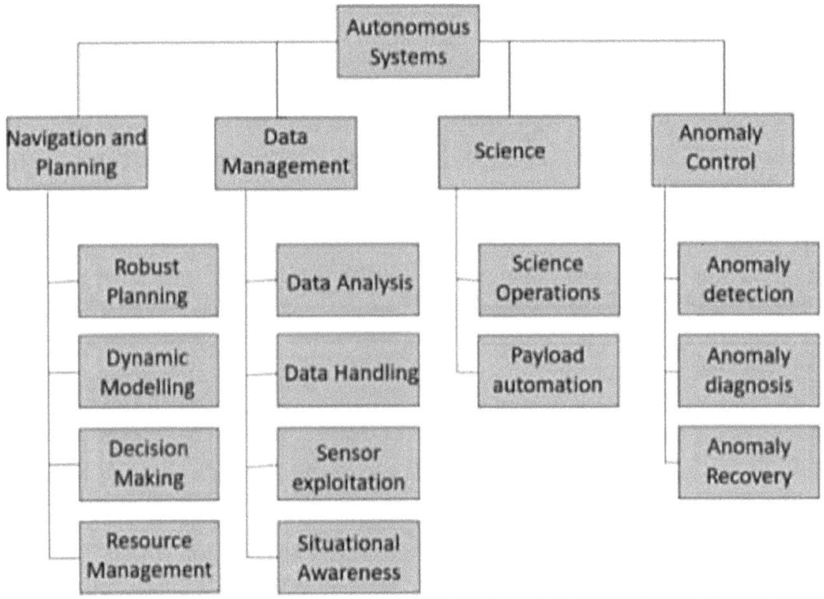

FIGURE 4.4 Future applications of autonomous system.

To prevent misuse, autonomous systems should only be used for the purpose for which they were intended. They should also be examined to identify any detrimental effects on people or communities. Policymakers must ensure that laws cover machines, software, and algorithms that power them as the nature of work and the role of humans in it change. Above all, humans must continue to be in charge of and benefit from machines. Autonomous systems may eventually enable firms to do more than just produce goods and services. An Industry 5.0 research and innovation initiative with a shared value business model has been identified by the EU. The production method prioritizes the welfare of the worker and makes use of cutting-edge technology like 5G and other digital technologies. Although major shifts in technology are rarely caused by a single incident, perhaps historians in the future will view the Suez blockade as a sign of the coming era of autonomous systems. A seismic change is currently taking place, and communications technology is the cause.

4.7 CONCLUSION

An autonomous system is one that can operate for an extended period of time without human intervention and accomplish a specified set of goals in a changing environment by gathering information about the environment. The two examples are autonomous mobile robots (AMRs) and driverless autos. An autonomous system needs to sense its surroundings and monitor its location and state at all times. It must establish a plan and decide what action to do next. It should avoid potentially dangerous circumstances that endanger human safety or the autonomous system itself, and it should only take action when it is safe to do so. The application of autonomous systems is becoming rapid with the incorporation of technologies like DevOps and AI/ML. We see that the use of autonomous systems varies from advanced sensors to autonomous drones and self-driving cars. We learn that it can be adapted beyond the research field and can be implemented in daily human control through the use of advanced sensors and can be used in high-tech navigation systems. The subject of autonomous systems is still budding; thus, the future of the technology is very vast and will develop at a much higher rate in the coming years. Developers will communicate with autonomous systems in the future through a software platform that abstracts the hardware, the sensors, and pushes the interface to a higher level of software services via an API. Developers will no longer need to work with a particular LiDAR, radar, or camera that is in use—instead, they will be able to only request information via high-level services provided with APIs. Similar abstractions will be used for connectivity, making it easier for developers to establish connections within the car, with other vehicles, and with the cloud without needing a thorough expertise of a particular communication technology.

KEYWORDS

- **artificial intelligence**
- **autonomous mobile robots**
- **autonomous system**
- **communication technology**
- **computer vision**
- **intelligent system**
- **IoT devices**

REFERENCES

1. Krishna, M., Neelima, M., Mane, H., & Matcha, V., (2018). *Image Identification Using Neural Networks, 7*, 614. 10.14419/ijet.v7i2.7.10892.
2. Huang, G. B., Zhu, Q. Y., & Siew, C. K., (2006). Extreme learning machine: Theory and applications. *Neurocomputing, 70*, 489–501.
3. Nguyen, G., et al., (2019). ML and DL frameworks and libraries for substantial and ample data mining: A survey. *Artif. Intell. Rev., 52*, 77–124.
4. Kshitiz, K., et al., (2017). Detecting hate speech and insults on social commentary using NLP and machine learning. *International Journal of Engineering Technology Science and Research 4*(12), 279–285.
5. Kumar, S., Dubey, S., & Gupta, P., (2015). Auto-selection and management of dynamic SGA parameters in RDBMS. In: *2015 2nd International Conference on Computing for Sustainable Global Development (INDIACom)* (pp. 1763–1768).
6. Niharika, S., Manish, P., Piyush, C., & Tanupriya, C., (2020). An optimized self-learning algorithm for autonomous navigation in an unknown environment. *Journal of Green Engineering*, 3315–3334.
7. Jing, T., Boglarka, V., Erika, T., Palya, F., Gabor, M. S., William, E. S., & DeBuc, D. C., (2016). Performance evaluation of automated segmentation software on optical coherence tomography volume data. *Journal of Biophotonics, 9*(5), 478–489.
8. Niharika, S., Manish, P., & Piyush, C., (2019). An efficient path planning algorithm for networked robots using modified optimization algorithm. *IJITEE*.
9. Hitesh, K. S., Anuj, K., Sangeeta, P., & Mangey, R., (2022). Artificial intelligence, blockchain and IoT for smart healthcare. In: *Artificial Intelligence, Blockchain and IoT for Smart Healthcare* (pp. i–xvi). River Publishers.
10. Hitesh, K. S., Anuj, K., Sangeeta, P., & Mangey, R., (2022). Introduction to smart healthcare and telemedicine systems. In: *Artificial Intelligence, Blockchain and IoT for Smart Healthcare* (pp. 1–12). River Publishers.

11. Hitesh, K. S., Anuj, K., Sangeeta, P., & Mangey, R., (2022). Advanced technologies involved in smart healthcare and telemedicine systems. In: *Artificial Intelligence, Blockchain and IoT for Smart Healthcare* (pp. 13–24). River Publishers.
12. Hitesh, K. S., Anuj, K., Sangeeta, P., & Mangey, R., (2022). 3 role of artificial intelligence, IoT and blockchain in smart healthcare. In: *Artificial Intelligence, Blockchain and IoT for Smart Healthcare* (pp. 25–36). River Publishers.
13. Ronald, K., & Barbara, E. K. K., (2013). The prevalence of age-related eye diseases and visual impairment in aging: Current estimates. *Investigative Ophthalmology & Visual Science, 54*(14).
14. Biswas, R., et al., (2012). A framework for automated database tuning using dynamic SGA parameters and basic operating system utilities. *Database Systems Journal, III*(4).
15. Sharma, H. K., (2013). E-COCOMO: the extended cost constructive model for cleanroom software engineering. *Database Systems Journal, 4*(4), 3–11.
16. Abadi, M., Agarwal, A., Barham, P., Brevdo, E., Chen, Z., Citro, C., ... & Zheng, X. (2016). Tensorflow: Large-scale machine learning on heterogeneous distributed systems. arXiv preprint arXiv:1603.04467.

CHAPTER 5

Comprehensive Study of Comfort and Motion Planning for Highway Autonomous Driving

SONALI VYAS and SHAURYA GUPTA

School of Computer Science, UPES, Dehradun, Uttarakhand, India

ABSTRACT

The prompt expansion in terms of autonomous vehicles has portrayed an efficient competent transference system. In terms of this concept implementation, considerable efforts are made to upscale the technical adeptness of autonomous vehicles. Self-driving automobiles will be a reality soon, as some of the major locomotive corporations have proclaimed that they will be selling their version of driving computerization modes in the early 2020s. The above-discussed technology has its significant disagreements in varied folds. However, technology in autonomous vehicles is still improving in terms of making individuals' lives much safer in terms of transportation, comfort, convenience, and lastly being eco-friendly. Nonetheless, human ease in the case of autonomous vehicles is under a transition phase and requires major technological uplift. Varied significant dynamics of comfort include sitting position, quivering, apart from noise. Some of the psychological factors leave a grander effect when human ease is considered. In spite of prevailing studies in terms of human ease besides some of the physiological signals which are studied in the case of automated driving contexts, and these are determined by the scale of human comfort level in autonomous vehicles which are detected with the help of these physiological

Advances in Autonomous Navigation Through Intelligent Technologies.
Niharika Singh, Thipendra P. Singh, Brian Azzopardi (Eds.)

signals. Autonomous vehicles vary from manually functioned automobiles in terms of the capability to extemporize as a human driver would do in certain circumstances. Ascertaining events that vehicles or automobiles encounter during their operation, so as to enable the decision-making process regarding the conclusions as how vehicles respond in addressing events of such types. The imminent phase of metropolitan transference surroundings is under transformation with the outline of self-sufficient automobiles amongst the networks. The design and development of such automobiles are employed in the interiors, which include closed and controlled testbeds responsible for simulating the metropolitan atmosphere. While various operations of such testbeds are performed, they consider the specific regulation of transportation operation. Motion planning indicates path formation besides decision-making in terms of traffic administration. Highway traffic situations face challenges like high-speed moving vehicles besides insignificant curvature roads, with definite driver guidelines, under a constrained environmental framework. There are numerous challenges like lane transformation, hindrance circumvention. Varied mechanisms in terms of motion planning, features, apart from their applications to highway driving are reviewed besides forthcoming challenges.

5.1 INTRODUCTION

Considering the last decade, the expansion of self-governing automobiles is spreading universally amongst various nations. The locomotive sector stands as one of the most hopeful progressions in terms of motorized manufacturing and its associated research. The first contributions in terms of driverless cars were by Houdina-Radio Control Company, later which was then effectively continued with its demonstration with the help of controlled radio signals which were being propelled by sprawling motor automobile [1]. Considering the period of the 1990s, investigation organizations in addition to locomotive industrialists partnered for frequent self-directed driving projects which were financed by the European EUREKA Prometheus program [56] implementing the concept of twin cars [2] apart from the concept of computerized highway platooning [3–5]. Currently, many industrial progressions in incomplete self-sufficient vehicles are in use with the aid of Advanced Driving Assistance Systems (ADAS). Concerning the freeway into consideration, main technologies concerns regarding longitudinal, lateral ease besides safety inclusive of cruise regulator, smart swiftness reworking, road assist. Although interrelating with difficulties, crash evasion systems moreover advise the chauffeur regarding a forthcoming collision which can be avoided by lane change assist action, which in turn is making the driver and car safe, varied other techniques including adaptive cruise control [6]. In many developed countries, including American states, laws have been embraced regarding the testing of self-directed automobiles in urban traffic management scenarios. The use and implementation of ADAS [7], which have been endorsed in computerized driving systems, are in accordance with the United Nations automobile conventions act [8]. With considerable improvement in the locomotive area, a customary cataloging for self-sufficient automobiles enabled with six-level systems [9]. Contestant teams, according to Buehler et al. [10] and Broggi et al. [11], prospered in circumnavigating a metropolitan situation, besides implementing the PROUD project [12] in 2013, which included a diverse transportation dynamic path through Parma, Italy. The aim of this analysis involves a discussion of a catalog of motion-oriented forecasting procedures for driving autonomous automobiles on the highway.

5.2 LITERATURE REVIEW

In the year 2011, varied European projects in Resende et al. [13]; Gühnemann et al. [14]; Alessandrini et al. [15] have been effectively established in terms of freeway besides metropolitan driving. In the year 2012, thoroughfare driving in platooning was attained by some of the project implementation [16, 17]. In 2016, authors discussed [18] assimilation, juncture, besides handling of disaster transportation situations. The robotics literature apart from numerous reviewing works which covers motion planning strategies [19–21] and includes precise use case on self-directed driving [22, 23, 55]. Additionally, thoroughfares involve the environment wherein chauffeurs or car drivers would be confidentially driving in a complete self-directed mode [24]. Autonomous vehicles stand at the brink of an insurrection in terms of the transportation system. The benefits of these kinds of vehicles are many which include relieving the drivers from lethargic or unfocused driving, augmented security, optimized energy consumption, and pollution [25, 26]. Notwithstanding the wonderful efforts given by technical experts in the enhancement of autonomous vehicle security and competence [27, 28]. A proper platform is required to conduct studies in terms of human comfort in autonomous vehicles. Any driving simulator that provides a precise simulated situation will offer a feasible solution for the study of such kinds. Numerous lessons considering human factors have shown the potential use of driving simulators as a substitute for real vehicles, focusing on human-centered studies. Additionally, driving simulators, available in both low and high-cost configurations, have been proven to generate feelings for the active participants of the study. Peterson et al. [29] focused on an automobile driving simulator equipped with a motion system that provided a comfortable ride, based on a comparative study between real and simulated test rides. Peterson et al. [30] focused on drivers' inclinations, apprehensions, including mental pressure and workload during autonomous vehicle rides, facilitated by a full-scale driving simulator. Ciloglu et al. [31]; Singh et al. [32] discussed a simulator including a screen, seat, navigation wheel, and gearshifts to examine the final impact of peril [33], as well as situational responsiveness [34], of drivers concerning the assistance system. In different readings, ease and uneasiness are understood as either different sides of the same construct [35, 36] or as two different constructs, each with significant aspects [37, 38]. The basic definition of comfort has been defined in various ways by different authors, as stated by authors De Vos et al. [39], which are defined in the following ways:

- Comfort is personally well-defined by each and every occupant, therefore inconsistency between individuals can be estimated.
- Comfort is always predisposed by varied influences.
- Comfort is subjective by an individual reaction to the environs and situations.

Before the swift expansion of autonomous vehicles, anthropological studies were primarily fixated majorly only on the physical comfort aspects. It included factors like tremor [40], noise [41], thermal comfort [42], air quality [43], etc. The primary negotiations regarding humanoid comfort in the case of autonomous vehicles concentrated on specific vehicular parameter controls. Jang et al. [44] discussed a semi-automated driving scenario with some variation factors like time progression and driver comfort. Nilsson et al.'s [45] study focused on the comfort of autonomous vehicles compared to their predecessor ones, like dispossession or lack of human driver's control on the vehicle. This included some psychosomatic features like the apparent level of safety and security, which are primarily very important aspects of humanoid ease whenever autonomous vehicles are considered. The psychological research by Song et al. [46] has observed the efficacy of using dissimilar machine learning approaches to help in identifying dissimilar categories of human sentiments with the help of physiological signals. Mouhagir et al. [47] discussed two approaches: support vector machine and Fisher discriminant, which helped in identifying individuals' emotions such as hilarity, satisfaction, repulsion, anxiety, neutrality, and unhappiness using various physiological signals including blood pulse volume, electrodermal events, skin heat, and credit significances for different emotion types. The results were excellent, with an accuracy of around 92%. Wang et al. [48] attempted to use various mathematical and machine learning techniques to analyze differences between different emotion types based on various physiological signals. The study successfully identified variations in emotional responses, with discriminant function analysis serving as an emotion recognizer. When studying human comfort, physiological indicators emerge as the most reliable markers. The effectiveness of utilizing multiple physiological indicators determines anthropometric thermal comfort calculation, extensively discussed by Katal et al. [49]. Aradi et al. [50] discussed EEG and EDA in the context of human comfort analysis during cycling. Qureshi et al. [51] explored correlations between human discomfort, facial expressions, and biological cues such as heart rate, pupil dilation, eye blinking, and EDA in the context of automated driving scenarios.

5.3 MOTION PLANNING SCHEME

A universal generalization of a categorized outline of self-directed automobile is discussed. In the case of motion strategy, the input data block is responsible for displaying data in terms of the ego vehicle, impediments' performances, and substructure portrayal, which is obtained from the discernment, localization, and communication blocks. It is immaterial how the information has been composed, but qualities like measures, uncertainties, and trustworthiness are taken into account. Therefore, motion strategy among perception, localization, and communication are discussed in detail by Van et al. [12]. This constituent accomplishes sensor information and delivers an awareness map with the following:

- Obstacles;
- Lanes;
- Traffic;
- Road;
- Ego information.

Strategy block informs character awareness besides localization of existing and forthcoming motion in a closed-loop system whenever considering a perception of communication. Apparently, it also conveys existing and forthcoming ego motion objectives to the surrounding. On the contrary, the control block gets input from reference motion, which is absolutely decided with the help of motion approach, and then it acts on actuators to facilitate movement of the ego vehicle. Information generated from sensors is responsible for productive outputs, which deliver an updated ego state vector. Furthermore, when dealing with semi-autonomous automobiles, the driver will interact with low levels of automation. In certain cases, communication between the driver and the motion strategy block, and vice versa, is quite essential in terms of warning the driver regarding motion decisions apart from his or her intent to drive the vehicle. The motion strategy is very much responsible for deciding the motion based on criteria categorized into three parts:

- Strategical that involves formation;
- Premeditated which involves direction; and
- Operative mechanism.

The architecture discussed by Van et al. [12] talks more about the above three layers. The strategic part is often divided into subparts, one designated for communicative forecasting and the other for motion generation [27]. The five main utilities in the case of motion strategy hierarchy are as follows:

- Way forecasting;
- Forecast;
- Conclusion making;
- Cohort; and
- Distortion.

The space and time horizons for each and every motion planning functions is being abridged in Table 5.1.

TABLE 5.1 Space and Time Horizons for Key Utilities

Utilities	Space	Time
Route planning	Greater than 100 m	Greater than 1 min and below 1 hour
Prediction	Greater than 1 m and below 100 m	Greater than 1 sec and below 1 min
Decision making	Greater than 10 m and below 100 m	Greater than 1 sec and below 1 min
Generation	Greater than 10 m and below 100 m	Greater than 1 sec and below 1 min
Deformation	Greater than 0.5 m and below 10 m	Greater than 10 ms and below 1s

Source: Claussmann et al. [52].

The first function, i.e., the route planner, is a trip scheduler which is responsible for providing overall long-term forecasting via the road network starting from the preliminary location to the driver's anticipated terminus. The second function, the prediction step, is accountable for storing the existing and momentous dynamics data which is used in predicting the dynamics of all sorts. Highway substructure, direction track, driving guidelines, track design development are typically inferred from map info. The acute chore of the forecasting function involves the forecast of obstacles of any sorts. Though it is very much needed for motion planning [29]. The essential purposes of motion planning scope are:

- Generation;
- Deformation.

Extrapolative forecasting is focused on three objectives:

- Danger assessment;
- Benchmarks minimization;
- Constraint compliance.

Above factors are involved in selecting the finest resolution for path generation as it engenders a set of motions which, in turn, helps in constructing decisions on behavioral motion.

5.4 HIGHWAY MOTION PLANNING CONSIDERATION

Considering the scenario in terms of highway motion, it is assumed that all vehicles are sensor-equipped. Taking a single vehicle into consideration, all others are treated as obstacles to that vehicle, wherein the kinematics of any vehicle is signified by states which include its locus and positioning, including the phase derivatives like point, speediness, and acceleration, which is termed as the configuration space. The scheme of autonomous ground vehicle is being discussed in Figure 5.1. The arrangement or configuration space, apart from the evolution space, is generally separated into three subspaces:

- Collision space wherein ego vehicle bumps with impediments on its way;
- Uncertain space wherein probability exists for ego vehicle to face a crash; and
- Free space in which there is no smash at all.

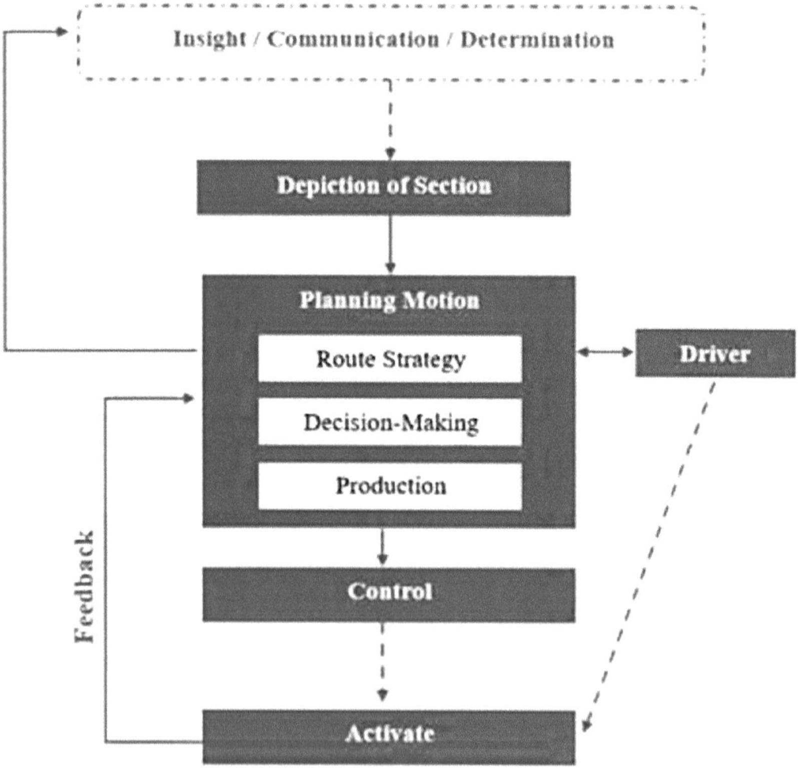

FIGURE 5.1 Scheme of autonomous ground vehicle.

Source: Claussmann et al. [52].

Motion refers to moreover a permitted space inclusive spatial geometric zones, geometric waypoints, trajectory which describes sequence of spatiotemporal states. Lastly, the prediction horizon signifies space or time horizon.

5.5 FREEWAY DRIVING SPECIFICATIONS

Motion development and planning practices exceedingly depend on use cases, generally deliberations for thoroughfare driving. The shape of roads is constructed in straight lines, circles with minor curves. Considering a nominal state, it consists of motor automobiles and adhering to driving guidelines. Consequently, traffic activities are very much uniform in terms of metropolitan driving. Singh et al. [32], insignificant non-exclusive circumstances are being discussed in a thoroughfare voyage:

1. **Track Possession:** The safety space which is in front and behind of the ego vehicle is certain. The longitudinal verdict ensures the anticipated speed, while the lateral verdict helps in keeping the ego route within the limit.
2. **Vehicle Succeeding:** In order to maintain its longitudinal safety expanses, the ego automobile necessarily should trail the front ego lane automobile.
3. **Lane Changing:** The judgment generally depends on varied hindrance constrictions; motion planner ensures that the target lane space is quite appropriate apart from the speed is also very much fine in order to keep the ego vehicle in a quite safe state.
4. **Lateral-Most Path Shifting:** In order to confirm flawless transportation movement, many driving guidelines require leftmost or rightmost driving conditions. Henceforth, the ego automobile continuously changes lanes until its lane, which is the lateral most, is set.
5. **Merging:** Highway driving involves a set of two actions at a time, i.e., incoming or departing the freeway, besides merging two lanes of automobiles into one. The ego automobile should be able to

acclimate to varied longitudinal and lateral speeds and distances to facilitate easy traffic movement.

6. **Highway Toll:** In this case, merging of vehicles in a fictitious path in expectation of a toll lane demarcation results in a gradual decrease of speed until the final stopping of the vehicle, while the decision to leave the toll lane involves acceleration of the reference speed of all vehicles in addition to merging into the final or usual traffic lane.

Autonomous vehicles are always considered in terms of platooning, and its conformation reduces the distances between automobiles while increasing the capability of highways. When considering the case of motion planning policy in platoons, it should be quite tough in terms of distinct automobiles in case of even acceleration, besides providing stability [3]. The key transformations amid freeway and metropolitan driving involve a sturdier emphasis on only the forward road direction, while urban driving encompasses a closer look in almost every direction. Thoroughfare automobile subtleties are quite naive in terms of:

- Low angle of turn;
- No back reversing;
- Least decelerating;
- Maintaining a constant and optimum speed.

So, even if there are fewer threats, the threats due to higher speed are quite stronger. Moreover, the greater speeds entail poorer sensor capabilities. Lastly, less traffic ensures a steadier situation. Therefore, algorithms that consider all specificities in a real-time environment will always be preferred in terms of practical application on thoroughfares.

5.6 CONSTRAINTS ON HIGHWAY DRIVING

For algorithmic conditions some of the precise freeway planning constrictions are considered which are as follows:

- High dependability;
- Safety;
- Low computation time desired.

Contrary to the atmosphere's safety constrictions, considering driving rules, avoiding impacted collision, and these form the hard constrictions as they are very much indispensable in terms of independent driving recognition. For a driver, it becomes very essential to consider ride optimization constrictions for:

- Time minimization;
- Distance;
- Energy consumption;
- Comfort intensification.

Above are the soft constraints which can be taken for ease. Whereas some of the possibility restrictions depend on kinematic limitations of the automobile. Finally, the selection of the automobile's model helps in handling constrictions which further encourages algorithmic complication with more degrees of freedom, and therefore a more intricate model solver is used. Degrees of freedom stand as an essential design constraint in terms of motion planning in addition to control architecture, which is discussed in guaranteeing a very harmless passable motion and consistency.

5.7 DESCRIPTION OF ANTHROPOLOGICAL EASE IN AUTONOMOUS VEHICLES

Personal ease is described in a variety of customs, as per studies conducted by Nor et al. [36]. Ease plus uneasiness instigated by pulsations in a vehicle was construed as a single continuation with two conclusions representing the most comfortable and uncomfortable in the study by Oborne. Although Singh et al. [3] imitated the ease and uneasiness of a person sitting on a chair as two independent entities. In this description, comfort of a person, in an autonomous vehicle is well explained as a basic structure having one end depicting discomfort and the other showing discomfort. Factors associated with human comfort were restricted to vehicle behavior. Only influences from vehicle behavior to human comfort are discussed here. We have described relief as relief from feelings of discomfort because it has been found to have a profound effect on a person's comfort. Nilsson et al. [45], luxury and discomfort as two independent builders were discussed. Although low levels of discomfort were initiated to be consistent with the rest of the array of ease levels. This has shown issues linked to luxury befitted inconsequential in existence of discomfort-related aspects. Here, we deliberated variations in human comfort related to motor behavior. Jang et al. [44] discussed loss of control which exposes the autonomous vehicle rider to novel significant features associated with uneasiness. Considering the case of conventional public transportation where behavior is not easy to handle, a passenger using AVs may feel uncomfortable because of losing control or encountering dangerous or unusual driving behavior. Research by Katal et al. [49] also recommends that awareness of AV driving style may play a significant part in an individual's fondness for autonomous vehicles. Diverse views about driving methods of AVs make a big difference in personal ease between conventional automobiles and autonomous vehicles. Despite the fact that the comfort features of a person in a traditional vehicle will have an impact on the autonomous vehicle. The study focuses on unique aspects of AVs and limits the range of personal comfort related to behavioral features for AVs.

5.8 CONCLUSION AND FUTURE SCOPE

A study of the content of this chapter shows that while the development of AV technology is predictable to be noteworthy, laws and policies are not yet developed adequately to address potential issues. AVs' limitations regarding the identification of distinctions and the law of spirit can be identified. Numerous ideas regarding ethical deliberations probable operators are found in the answers to the survey. Central to them is the perceived balance between amenability amid regulation besides variation to the surrounding atmosphere. The consistent balance previous actual choice of actions remains quite evident in a number of cases. One outstanding example of the difference is that AVs can exceed the legal limit based on the fact that the current situation is critical to security. In spite of any potential implications for the findings, the boundaries of the study should be considered. First, the size of the sample observed remains comparatively lesser because the key purpose of this study was to closely observe the problem and identify key features and conflicts. Finally, the sample was restricted mainly to the city inhabitants. As this will not bring a big change if the city does not have a well-organized and comprehensive public transport network. The above limits indicate the upcoming research directions. Forthcoming surveys will comprise a huge sample size which needs to be discussed for better distribution under conscious situations. Further cases deal with circumstances wherein autonomous vehicles will be used in emergency and disaster situations.

KEYWORDS

- **autonomous vehicles**
- **driver assistance system**
- **human ease recognition**
- **motion and path planning**
- **physiological signal**
- **virtual environment**

REFERENCES

1. Bird, S. E., (1992). *For Enquiring Minds: A Cultural Study of Supermarket Tabloids*. Univ. Of Tennessee Press.
2. Ulmer, B., (1994). Vita ii-active collision avoidance in real traffic. In: *Proceedings of the Intelligent Vehicles' 94 Symposium* (pp. 1–6). IEEE.
3. Singh, N., Prateek, M., & Chauhan, P., (2019). An efficient path planning algorithm for networked robots using modified optimization algorithm. In: *International Journal of Innovative Technology and Exploring Engineering* (Vol. 8, No. 12, pp. 2738–2743). Blue Eyes Intelligence Engineering and Sciences Engineering and Sciences Publication – BEIESP.
4. Broggi, A., Bertozzi, M., Fascioli, A., Bianco, C. G. L., & Piazzi, A., (1999). The ARGO autonomous vehicle's vision and control systems. *International Journal of Intelligent Control and Systems, 3*(4), 409–441.
5. Rajamani, R., Choi, S. B., Law, B. K., Hedrick, J. K., Prohaska, R., & Kretz, P., (2000). Design and experimental implementation of longitudinal control for a platoon of automated vehicles. *J. Dyn. Sys. Meas., Control, 122*(3), 470–476.
6. Zechmeister, H. G., Dullinger, S., Hohenwallner, D., Riss, A., Hanus-Illnar, A., & Scharf, S., (2006). Pilot study on road traffic emissions (PAHs, heavy metals) measured by using mosses in a tunnel experiment in Vienna, Austria. *Environmental Science and Pollution Research, 13*(6), 398–405.
7. Hasenjäger, M., & Wersing, H., (2017). Personalization in advanced driver assistance systems and autonomous vehicles: A review. In: *2017 IEEE 20th International Conference on Intelligent Transportation Systems (ITSC)* (pp. 1–7). IEEE.

8. Poggenhans, F., Pauls, J. H., Janosovits, J., Orf, S., Naumann, M., Kuhnt, F., & Mayr, M., (2018). Lanelet2: A high-definition map framework for the future of automated driving. In: *2018 21st International Conference on Intelligent Transportation Systems (ITSC)* (pp. 1672–1679). IEEE.
9. Körber, M., Baseler, E., & Bengler, K., (2018). Introduction matters: Manipulating trust in automation and reliance in automated driving. *Applied Ergonomics, 66*, 18–31.
10. Buehler, M., Iagnemma, K., & Singh, S., (2009). *The DARPA Urban Challenge: Autonomous Vehicles in City Traffic* (Vol. 56). Springer.
11. Broggi, A., Buzzoni, M., Debattisti, S., Grisleri, P., Laghi, M. C., Medici, P., & Versari, P., (2013). Extensive tests of autonomous driving technologies. *IEEE Transactions on Intelligent Transportation Systems, 14*(3), 1403–1415.
12. Van, S. I., Welsh, R., Backer-Grondahl, A., Hoedemaeker, M., Lotan, T., Morris, A., & Winkelbauer, M., (2011). *Towards a Large Scale European Naturalistic Driving Study: Final Report of Prologue: Deliverable D4,* 2.
13. Resende, P., Pollard, E., Li, H., & Nashashibi, F., (2013). *ABV-A Low Speed Automation Project to Study the Technical Feasibility of Fully Automated Driving.* In Workshop PAMM.
14. Gühnemann, A., Pfaffenbichler, P., Shepherd, S., & Emberger, G., (2018). Simulating transport and societal effects of automated vehicles. In: *36th International Conference of the System Dynamics Society*. Reykjavík, Iceland.
15. Alessandrini, A., Holguin, C., & Parent, M., (2018). The certification approach for ARTS. *Implementing Automated Road Transport Systems in Urban Settings*, 266.
16. Ziegler, J., Bender, P., Schreiber, M., Lategahn, H., Strauss, T., Stiller, C., & Zeeb, E., (2014). Making bertha drive—An autonomous journey on a historic route. *IEEE Intelligent Transportation Systems Magazine, 6*(2), 8–20.
17. Englund, C., Chen, L., Ploeg, J., Semsar-Kazerooni, E., Voronov, A., Bengtsson, H. H., & Didoff, J., (2016). The grand cooperative driving challenge 2016: Boosting the introduction of cooperative automated vehicles. *IEEE Wireless Communications, 23*(4), 146–152.
18. Latombe, J. C., (2012). *Robot Motion Planning* (Vol. 124). Springer Science & Business Media.
19. Laumond, J. P., Sekhavat, S., & Lamiraux, F., (1998). Guidelines in nonholonomic motion planning for mobile robots. In: *Robot Motion Planning and Control* (pp. 1–53). Springer, Berlin, Heidelberg.
20. LaValle, S. M., (2006). *Planning Algorithms*. Cambridge University Press.
21. Katrakazas, C., Quddus, M., Chen, W. H., & Deka, L., (2015). Real-time motion planning methods for autonomous on-road driving: State-of-the-art and future research directions. *Transportation Research Part C: Emerging Technologies, 60*, 416–442.
22. González, D., Pérez, J., Milanés, V., & Nashashibi, F., (2015). A review of motion planning techniques for automated vehicles. *IEEE Transactions on Intelligent Transportation Systems, 17*(4), 1135–1145.
23. Paden, B., Čáp, M., Yong, S. Z., Yershov, D., & Frazzoli, E., (2016). A survey of motion planning and control techniques for self-driving urban vehicles. *IEEE Transactions on Intelligent Vehicles, 1*(1), 33–55.

24. Fernandes, P., & Nunes, U., (2012). Platooning with IVC-enabled autonomous vehicles: Strategies to mitigate communication delays, improve safety and traffic flow. *IEEE Transactions on Intelligent Transportation Systems, 13*(1), 91–106
25. Rödel, C., Stadler, S., Meschtscherjakov, A., & Tscheligi, M., (2014). Towards autonomous cars: The effect of autonomy levels on acceptance and user experience. In*: Proceedings of the 6th International Conference on Automotive User Interfaces and Interactive Vehicular Applications* (pp. 1–8).
26. Power, J. D., (2019). *Mobility Pipe Dreams? JD Power and Survey Monkey Uncover Shaky Consumer Confidence About the Future*. JD Power.
27. Bellem, H., Klüver, M., Schrauf, M., Schöner, H. P., Hecht, H., & Krems, J. F., (2017). Can we study autonomous driving comfort in moving-base driving simulators? A validation study. *Human Factors, 59*(3), 442–456.
28. Du, N., Haspiel, J., Zhang, Q., Tilbury, D., Pradhan, A. K., Yang, X. J., & Robert, Jr. L. P., (2019). Look who's talking now: Implications of AV's explanations on driver's trust, AV preference, anxiety and mental workload. *Transportation Research Part C: Emerging Technologies, 104*, 428–442.
29. Petersen, L., Zhao, H., Tilbury, D., Yang, X. J., & Robert, L., (2018). *The Influence of Risk on Driver's Trust in Semi-Autonomous Driving.*
30. Petersen, L., Robert, L., Yang, J., & Tilbury, D., (2019). Situational awareness, driver's trust in automated driving systems and secondary task performance. *SAE International Journal of Connected and Autonomous Vehicles, Forthcoming.*
31. Ciloglu, H., Alizadeh, M., Mohany, A., & Kishawy, H., (2015). Assessment of the whole body vibration exposure and the dynamic seat comfort in passenger aircraft. *International Journal of Industrial Ergonomics, 45,* 116–123.
32. Singh, N., Prateek, M., Chauhan, P., & Choudhury, T., (2020). An optimized self-learning algorithm for autonomous navigation in an unknown environment. *Journal of Green Engineering, 10*, 3315–3334.
33. De Looze, M. P., Kuijt-Evers, L. F., & Van, D. J. A. A. P., (2003). Sitting comfort and discomfort and the relationships with objective measures. *Ergonomics, 46*(10), 985–997.
34. Bishu, R. R., Hallbeck, M. S., Riley, M. W., & Stentz, T. L., (1991). Seating comfort and its relationship to spinal profile: A pilot study. *International Journal of Industrial Ergonomics, 8*(1), 89–101.
35. Parsons, K. C., Whitham, E. M., & Griffin, M. J., (1979). Six axis vehicle vibration and its effects on comfort. *Ergonomics, 22*(2), 211–225.
36. Nor, M. J. M., Fouladi, M. H., Nahvi, H., & Ariffin, A. K., (2008). Index for vehicle acoustical comfort inside a passenger car. *Applied Acoustics, 69*(4), 343–353.
37. Devonshire, J. M., & Sayer, J. R., (2005). Radiant heat and thermal comfort in vehicles. *Human Factors, 47*(4), 827–839.
38. Ormuž, K., & Muftić, O., (2004). Main ambient factors influencing passenger vehicle comfort. *Proceedings of 2nd International Ergonomics*. Stubičke Toplice, Zagreb, Croatia.
39. De Vos, A. P., Theeuwes, J., Hoekstra, W., & Coëmet, M. J., (1997). Behavioral aspects of automatic vehicle guidance: Relationship between headway and driver comfort. *Transportation Research Record, 1573*(1), 17–22.

40. Elbanhawi, M., Simic, M., & Jazar, R., (2015). In the passenger seat: investigating ride comfort measures in autonomous cars. *IEEE Intelligent Transportation Systems Magazine, 7*(3), 4–17.
41. Lisetti, C. L., & Nasoz, F., (2004). Using noninvasive wearable computers to recognize human emotions from physiological signals. *EURASIP Journal on Advances in Signal Processing, 2004*(11), 1–16.
42. Kim, J., & André, E., (2008). Emotion recognition based on physiological changes in music listening. *IEEE Transactions on Pattern Analysis and Machine Intelligence, 30*(12), 2067–2083.
43. Maaoui, C., & Pruski, A., (2010). Emotion recognition through physiological signals for human-machine communication. *Cutting Edge Robotics, 2010*(317–332), 11.
44. Jang, E. H., Park, B. J., Park, M. S., Kim, S. H., & Sohn, J. H., (2015). Analysis of physiological signals for recognition of boredom, pain, and surprise emotions. *Journal of Physiological Anthropology, 34*(1), 1–12.
45. Nilsson, J., Fredriksson, J., & Coelingh, E., (2017). Trajectory planning with miscellaneous safety critical zones. *IFAC-PapersOnLine, 50*(1), 9083–9088.
46. Song, X., Cao, H., & Huang, J., (2013). Vehicle path planning in various driving situations based on the elastic band theory for highway collision avoidance. *Proceedings of the Institution of Mechanical Engineers, Part D: Journal of Automobile Engineering, 227*(12), 1706–1722.
47. Mouhagir, H., Talj, R., Cherfaoui, V., Aioun, F., & Guillemard, F., (2016). Integrating safety distances with trajectory planning by modifying the occupancy grid for autonomous vehicle navigation. In: *2016 IEEE 19th International Conference on Intelligent Transportation Systems (ITSC)* (pp. 1114–1119). IEEE.
48. Wang, Q., & Ayalew, B., (2016). Obstacle filtering algorithm for control of an autonomous road vehicle in public highway traffic. In: *Dynamic Systems and Control Conference* (Vol. 50701, p. V002T24A009). American Society of Mechanical Engineers.
49. Katal, A., & Singh, N., (2022). Artificial neural network: Models, applications, and challenges. *Innovative Trends in Computational Intelligence*, 235–257.
50. Aradi, S., (2020). Survey of deep reinforcement learning for motion planning of autonomous vehicles. *IEEE Transactions on Intelligent Transportation Systems.*
51. Qureshi, A. H., Miao, Y., Simeonov, A., & Yip, M. C., (2020). Motion planning networks: Bridging the gap between learning-based and classical motion planners. *IEEE Transactions on Robotics, 37*(1), 48–66.
52. Claussmann, L., Revilloud, M., Gruyer, D., & Glaser, S., (2019). A review of motion planning for highway autonomous driving. *IEEE Transactions on Intelligent Transportation Systems, 21*(5), 1826–1848.
53. Stocker, A., & Shaheen, S., (2018). Shared automated mobility: early exploration and potential impacts. *Road Vehicle Automation, 4*, 125–139.
54. Litman, T., (2020). *Autonomous Vehicle Implementation Predictions: Implications for Transport Planning.*
55. Litman, T. (2017). *Autonomous Vehicle Implementation Predictions.*
56. Stöcker, C., Bennett, R., Nex, F., Gerke, M., & Zevenbergen, J. (2017). Review of the current state of UAV regulations. *Remote Sensing, 9*(5), 459.

CHAPTER 6

A Review on Self-Learning Routing Protocols for Autonomous Unmanned Aerial Networks (A-UAN)

HINA SHOUKAT,[1] MUHAMMAD YASEEN AYUB,[1]
MARIYA OUAISSA,[2] and INAM ULLAH KHAN3

[1]*Department of Computer Science, COMSATS University Islamabad, Attock Campus, Pakistan*

[2]*Moulay Ismail University, Meknes, Morocco*

[3]*King's College London, UK*

ABSTRACT

Unmanned aerial vehicles are the newly emerged technology which has applications in various fields of study. Remote-controlled UAVs are connected to satellites and ground-based stations. Therefore, routing protocols play an important role in effective communication within autonomous aerial networks. Due to the high mobility patterns of aerial vehicles, the deployment of routing protocols is quite difficult. The topological environment consists of ground-based IoT sensor nodes where UAVs collect information and send it to the base station. Moreover, the performance and efficiency of routing algorithms in FANETs are the optimal solution for better communication. Although the areas of flying ad hoc networks are in their infancy, researchers use traditional routing protocols that are not effective. This research chapter provides a review study on routing protocols related to unmanned aerial networks. In addition, challenges, trends, and future directions are also discussed in detail.

Advances in Autonomous Navigation Through Intelligent Technologies.
Niharika Singh, Thipendra P. Singh, Brian Azzopardi (Eds.)

6.1 INTRODUCTION

Unmanned aerial vehicles (UAVs) have already been deployed in many roles in both military and civil domains. UAVs are used to minimize human intervention, especially in dangerous areas where human life is at risk. For example, UAVs can be used in delivering medical aid in remote areas, monitoring volcanic eruptions, or monitoring other areas where human physical access is not possible, such as in combat zones and numerous other types of monitoring [9]. With the increased use of UAVs, it is essential to develop a secure and fast communication mechanism among drones so that they can provide uninterrupted service. Over the years, researchers have developed numerous routing protocols for UAVs which provide them a fast and efficient way of communicating with each other and the base station. There are several types of challenges faced by UAVs during flight. Due to high speed and rapid topological changes, UAVs sometimes lose connection, which may even result in a crash. To overcome this issue, researchers have worked on different routing protocols which can provide safe and secure communication between UAVs. Due to rapidly changing terrain and topology in UAV networks, it was difficult to set up a UAV network. UAV technology saw a massive boom with the development of Ad-hoc networks for UAVs, as in Ad-hoc networks there is no requirement of a pre-existing communication setup. In a UAV ad-hoc network, one drone known as the backbone UAV is responsible for acting as a bridge or gateway between the base station on the ground and other UAVs [17], as shown in Figure 6.1.

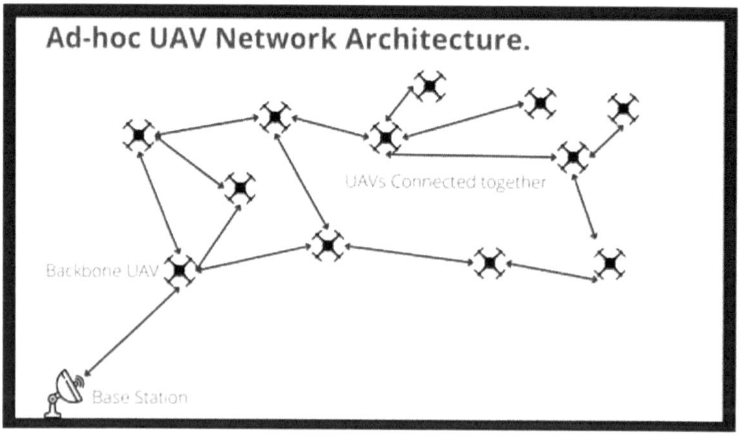

FIGURE 6.1 Ad-hoc UAV network.

As shown in Figure 6.1, the backbone UAV is the only UAV in the network that is connected to the base station. The backbone UAV has both short and long-range communication systems. Therefore, this approach significantly increases the communication range of the UAV network.

6.2 LITERATURE REVIEW

Integration of 5G networks in UAV networks has improved overall communication. FANETs have emerged from mobile ad hoc networks. The topological structure of MANETs is quite different compared to FANETs. There is a special need for novel routing protocols in the area of FANETs. Lightweight routing protocols are quite helpful and easy to deploy in UAV networks. Additionally, 5G communication has reduced end-to-end delay, and missions of UAVs are completed quickly. Implementing routing protocols is considered a tough job in the dynamic pattern of UAV networks. Mobility models enhance the ability of routing protocols in networks. Initially, traditional routing protocols like AOMDV, DSR, DSDV, ZRP, and M-DART are proposed for FANETs. However, these routing protocols have issues due to the continuous movement of UAVs [1]. UAVs in smart cities are a new concept that requires efficient communication. Aerial vehicles use IEEE 802.11 as the standard protocol for sharing information with the base station. Intelligent transportation methods need better routing protocols in FANETs. IoT is used to allow UAVs in smart cities for optimal communication using meta-heuristic routing. An improved sequencing heuristic DSDV protocol is designed, which shows better results in throughput, packet drop rate, and end-to-end delay. The nomadic mobility model is utilized to check the performance of UAVs [2]. A reinforcement-based routing protocol is formulated in FANETs. Also, wireless communication technology is used as a backbone [3]. The E-AntHocNet routing protocol is proposed, which is based on ant colony optimization. A novel energy stabilizing threshold is designed, which has improved overall network performance [4]. Hassan et al. have introduced the fisheye state routing protocol in the area of FANETs. For comparison, DSDV, AODV, TORA, and OLSR are used to evaluate the performance. Therefore, FSR in UAV-networks has shown better results, which have improved communication channels [5]. However, IoT networks have connectivity problems, which lead to miscommunication [6]. UAVs have high speed and specific direction; therefore, the AntHocNet routing protocol is quite effective for better communication. Moreover, a machine learning-based classifier decision tree is used to improve UAV-to-ground wireless communication links [7, 8].

Therefore, signal strength and routing protocols have vulnerabilities in aerial networks that need to be addressed properly by using various techniques. In the early days of FANETs, existing routing protocols that were mostly developed for Mobile Ad-hoc Networks (MANETs) were implemented into flying ad-hoc networks (FANETs) as shown in Table 6.1. Due to different characteristics in both Ad hoc networks, results were unsatisfactory.

TABLE 6.1 Routing Protocols Initially Used for FANETs

Type of Routing Protocol Initially Used for FANETs	References
Destination sequenced distance vector (DSDV)	[10]
Optimized link state routing (OLSR)	[11]
Ad hoc on-demand distance vector (AODV)	[12]
Dynamic source routing (DSR)	[13]
Temporally ordered routing algorithm (TORA)	[14]

FANET-specific routing protocols first started to appear in the year 2006 [15]. They were primarily based on Ad hoc On-demand Distance Vector (AODV) routing protocols and were proposed in Ad-hoc Routing Protocol for Aeronautical Mobile Ad hoc Networks (ARPAM) [16]. This was followed by a series of further advancements in developing advanced routing protocols that could be effectively used in FANETs. Table 6.2 shows the evolution of routing protocols used in FANETs since 2006.

These routing methods are old and require human supervision during UAV operation. For future use where UAVs will be used extensively in different domains to assist humans, new routing protocols need to be developed which can enable UAVs to perform autonomous flight missions.

6.3 UNMANNED AERIAL VEHICLES IN SMART CITIES

For the past few decades, the world's population has been increasing rapidly, and it is expected to double by the end of 2050. Due to this rapid population growth, many people are moving to cities in search of a better

TABLE 6.2 FANET Routing Protocols from 2006 to 2019

Year	FANET Routing Protocols Introduced
2006	ARPAM
2008	CAUN
2009	USMP
2010	DOLSR
2011	MPCA
2012	RGR, GPMOR, GHRAA
2013	P-OLSR, RTORA, AFP, A-GR
2014	ML-OLSR, LEBR, modified-RGR, XLinGO
2015	MENUM
2016	FGQPA, GCS-routing, DRS
2017	GPNC-SP, ADRP, BR-AODV, RARP, LEPR
2018	ECRNET, IMRL, improved OLSR-ETX, SARP, GeoSaw, GRP, NC-OLSR, SD-UAVA, JarmRout
2019	BICSF, NCMR

life and more amenities. To improve quality of life and efficiency, we are transitioning towards smart cities and self-managing machines. Unmanned aerial vehicles (UAVs) are particularly valuable for tasks such as delivery, infrastructure inspection, traffic monitoring, and environmental surveillance. Figure 6.2 illustrates FANETS in a smart city architecture. While UAVs were previously primarily used for military purposes, they are now employed in various sectors such as agriculture, civil applications, and military operations. In smart cities, UAVs can help manage traffic and provide emergency assistance during accidents or incidents. UAVs are also utilized in firefighting, detecting suspicious behavior, monitoring crowds, delivering parcels, and other daily activities. In agriculture, UAVs equipped with low-cost swarm drone technology can be used for plant protection, pesticide spraying, and combating issues like locust infestations [18].

6.4 MOBILITY MODELS FOR UAV-NETWORK

Unmanned Aerial Vehicles (UAVs) can have a wide range of applications in various types of environments. Depending on the environment they

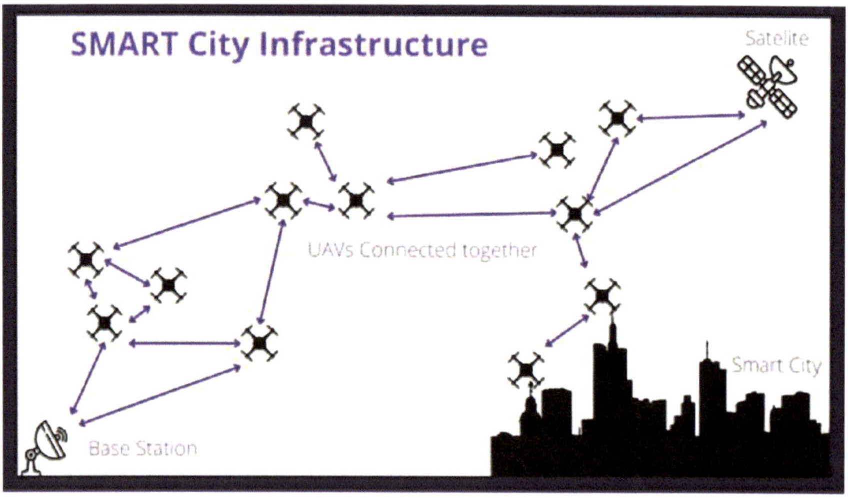

FIGURE 6.2 FANETs in smart city infrastructure.

are used in and the type of operation they are performing, their mobility models are different. Mobility models are designed for simulation environments that represent the changes in speed, trajectory, and acceleration of UAV nodes in a network. They play a vital role in the evaluation of any FANET design [19].

6.4.1 RANDOM WALK MOBILITY MODEL

The random walk mobility model is one of the simplest mobility models. It is a memoryless mobility model, meaning that no previous speed and location information is stored. The random walk mobility model selects a random speed and direction each time. After a fixed interval, a new random speed and direction is selected [20].

6.4.2 RANDOM WAYPOINT MOBILITY MODEL

The random waypoint mobility model is quite similar to the random walk mobility model with a few additions. The random waypoint model is based on choosing a random destination instead of a random direction. After each interval, there is a pause during which the node is completely stationary. Meanwhile, it selects its next random destination and in the next time interval, it reaches that random destination [21].

6.4.3 NOMADIC COMMUNITY MOBILITY MODEL

Nomadic Community mobility models are those mobility models in which more than one UAV moves from one location to another together around a reference point. In the Nomadic Community mobility model, some nodes may share the same location and collide, to avoid which some additional limitations are applied. This model is widely applied in agricultural and military domains where multi-UAVs are flying to perform an operation [22].

6.5 ROUTING PROTOCOLS FOR UAV-NETWORKS

Routing protocols are a set of rules that are followed by two devices to communicate with each other and to share information. It is the algorithm used to find the best path between the source and destination. UAVs also use some sort of routing protocol to communicate with each other or with the base station. Figure 6.3 depicts the best path found by the routing protocol for communication between a node and the base station.

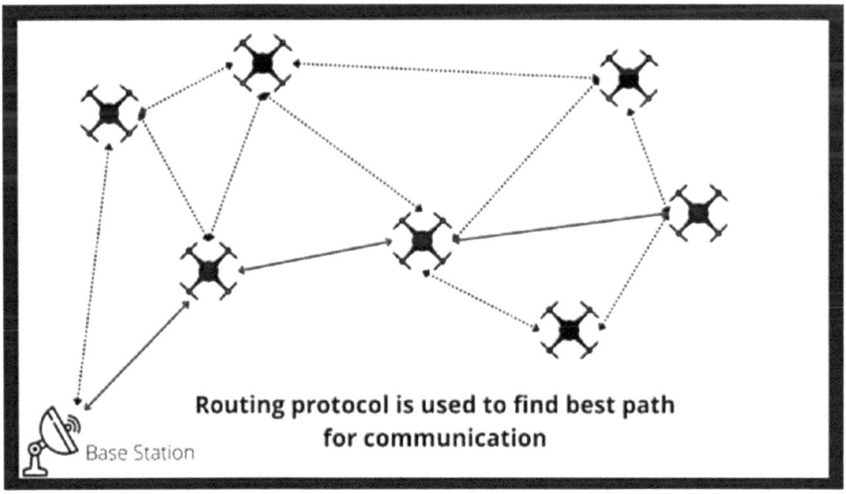

FIGURE 6.3 Routing protocols used to determine best path for communication.

6.5.1 DYNAMIC SOURCE ROUTING (DSR)

Dynamic Source Routing (DSR) is kind of AODV because of its route on-demand design, however instead of following a routing table it relies on source routing. It forms a self-configuring and self-organizing network. The source UAV broadcasts a Route Request to its neighbor UAVs. As there can be multiple nodes sending route requests in the network, to avoid any confusion the source UAV also adds a destination request ID with the route request.

6.5.2 LOW-ENERGY ADAPTIVE CLUSTERING HIERARCHY (LEACH)

Wireless UAV networks always face a challenge of being power limited. Thus, an efficient use of available battery is essential. The Low-Energy Adaptive Clustering Hierarchy (LEACH) routing protocol is basically used for effective power management. The LEACH protocol minimizes the number of messages transmitted to the sink by using multi-hop communication within a cluster, performing data fusion and data aggregation [24]. Figure 6.4 shows the architecture of the LEACH protocol.

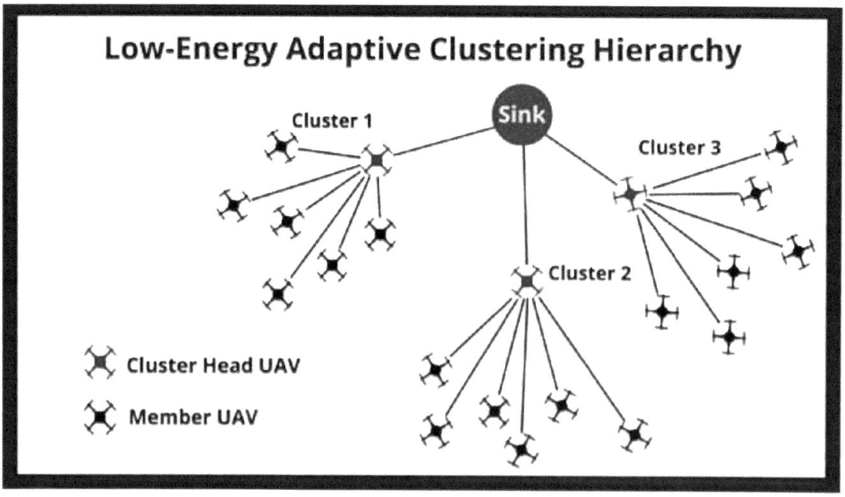

FIGURE 6.4 LEACH protocol architecture.

6.5.3 ZONE ROUTING PROTOCOL (ZRP)

Zone routing is based on the creation of different zones. A zone is created with a set of UAV nodes. Routing inside a zone can be done instantly as the source and destination UAV are present inside the same zone. This is called intra-zone routing. When the destination UAV is present in another zone and communication happens across a different zone, then it is called inter-zone routing [23] as depicted in Figure 6.5.

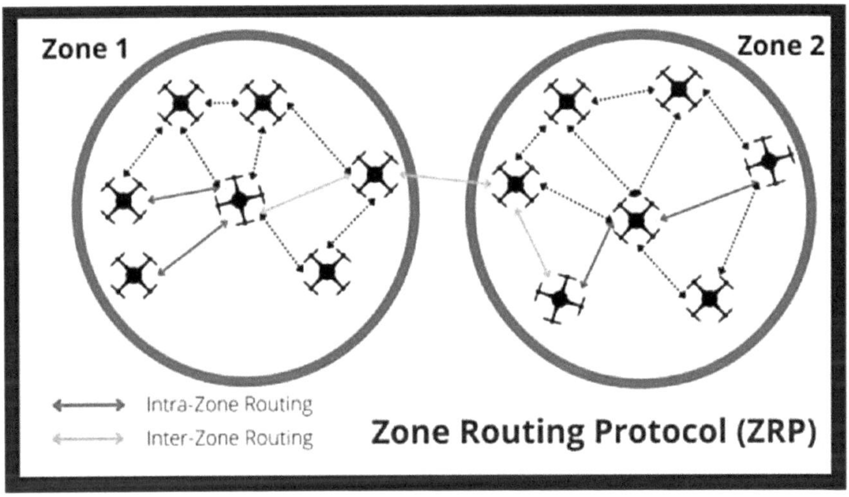

FIGURE 6.5 Zone routing protocol between two group of UAVs.

6.5.4 MULTI-PATH DYNAMIC ADDRESS ROUTING (M-DART)

M-DART is an extension of the DART protocol. It proactively discovers all available routes in a wireless UAV network. It uses a hop-by-hop routing approach and is based on the distance vector concept. M-DART stores all available routes to each sibling, unlike the Dynamic Address Routing protocol, which stores only one route for each sibling. There is no extra or additional control packet or signal required in M-DART. It uses the same number of entries in the routing update packet as in the Dynamic Address Routing protocol [25]. Figure 6.6 shows the routing update packet of both DART and M-DART.

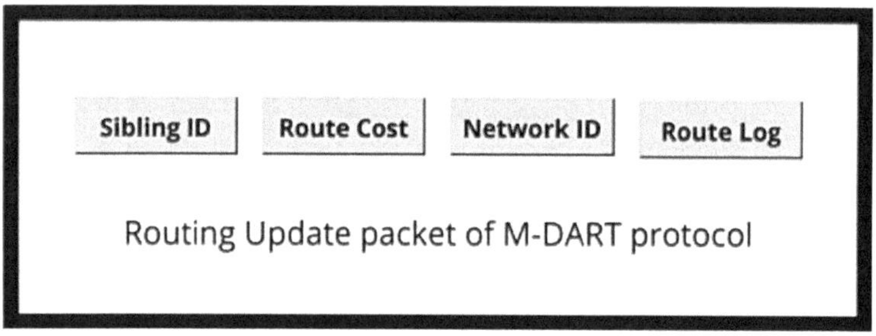

FIGURE 6.6 Routing update packet of DAR and M-DAR protocols.

6.5.5 OPTIMIZED LINK STATE ROUTING (OLSR)

OLSR algorithm is a table-driven link state algorithm. It is an optimized form of the classical pure link state protocol used for MANETs, tailored for wireless ad-hoc networks. It is very effective in large and dense FANETs. As the size of the network grows, some nodes start to disappear due to limited range. OLSR allows those nodes to communicate through intermediate nodes [26]. OLSR works on the concept of Multipoint relays (MPRs). During the flooding process, these multipoint relays (MPRs) are selected nodes which forward broadcast messages, reducing message overhead compared to the classical pure link state protocol. In OLSR, only selected MPRs can generate link state information, thus reducing the number of control messages in the network [27]. Figure 6.7 shows the architecture of the OLSR protocol.

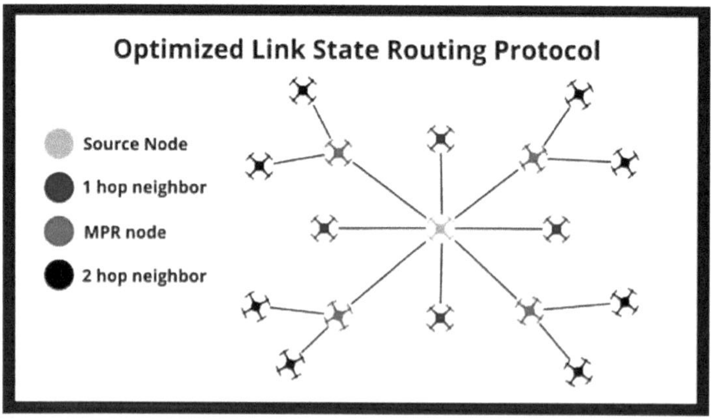

FIGURE 6.7 Architecture of optimized link state routing.

6.6 MACHINE LEARNING FOR UAVS

Machine Learning is the development of systems that enable a computer to perform human cognitive decisions. Machine learning basically tries to replicate human cognitive functions. A machine learning-enabled system can effectively make decisions in complex environments where traditional decision-making approaches start to lag behind or often fail. Depending on operating methods, Machine Learning is divided into two main categories, which are Supervised Learning and Unsupervised Learning as shown in Figure 6.8.

6.6.1 SUPERVISED LEARNING

Supervised learning algorithms are mainly classification algorithms with the end goal of learning how to classify one or more classes differently. In Supervised Learning, a labeled dataset is provided, and the machine learns to map input data to output classes during the training phase. Supervised learning algorithms include classification algorithms like Decision tree, SVM, Naïve Bayes, etc., and logistic regression [28].

6.6.2 UNSUPERVISED LEARNING

As the name Unsupervised suggests, these are types of algorithms which are not provided with labeled data during the training phase. These algorithms find certain patterns in data on their own and based on that, group them. An example of unsupervised learning is clustering [29].

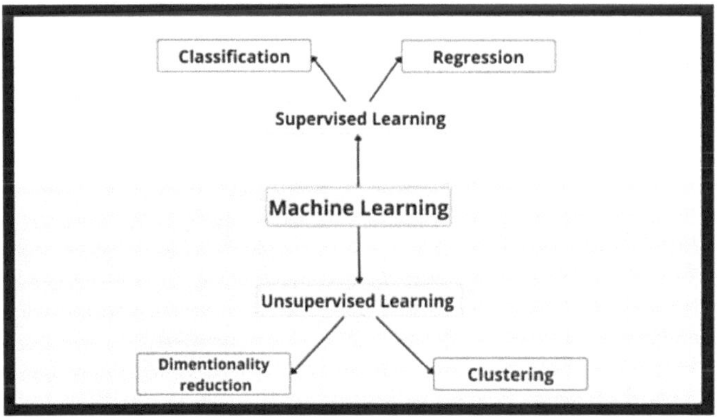

FIGURE 6.8 Types of machine learning.

6.7 UNMANNED AERIAL VEHICLES APPLICATIONS

UAVs have many applications in different domains. In FANETs, with the potential autonomous behavior of UAVs, human involvement is minimized. This has further increased the applications of Unmanned Aerial Vehicles such as surveillance/monitoring, environment sensing, relaying network, emergency medical aid, firefighting, military combat, etc. [30].

6.7.1 *RESCUE OPERATIONS USING UAVS*

UAVs can be used in various types of rescue missions from military to civil. Yeong, King, & Dol [31] discusses how UAVs are used in marine search and rescue missions. The main goal of search and rescue missions is to quickly find the rescue target in the shortest possible time. Any delay will reduce the chances of survival of victims. UAVs can also be used to deliver emergency medical aid to areas that are inaccessible by road. UAVs can also be used for the emergency supply of medical aid in remote areas where otherwise rapid response is not possible. UAVs can be quickly dispatched from the nearest hospital containing medical aid, and they can reach the destination quickly as depicted in Figure 6.9.

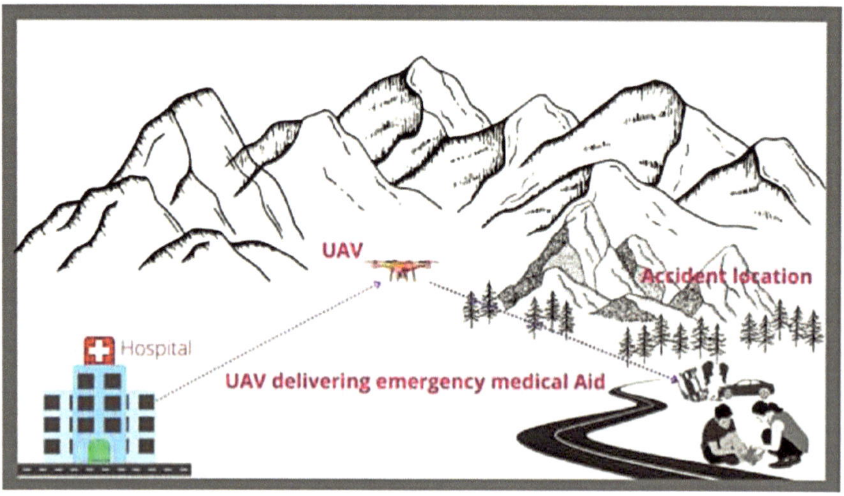

FIGURE 6.9 Emergency supply of medical aid using UAV.

6.7.2 UAVS IN DISASTER MANAGEMENT

UAVs can be widely employed in disaster management and monitoring. The quickest way of natural disaster assessment is through aerial monitoring. UAVs can effectively be used in monitoring natural disasters such as earthquakes, volcanoes, avalanches, landslides, tsunamis, flash floods, wildfires, etc. In such situations, response time is key to saving lives. UAVs can minimize response time and provide situational awareness [32]. During flash floods when road connectivity is badly affected, UAVs can be a game changer in damage analysis and coordinating proper relief operations. UAVs can also be helpful in pandemic control for monitoring people.

During disaster management, UAVs can also act as a standalone communication system. They can be used to re-establish the damaged communication structure in affected areas.

6.7.3 UAVS IN FOREST-BASED APPLICATIONS

UAVs can also be used in forestry and agriculture departments. UAVs can be very effective in helping to plant man-made forests. Rapid reforestation activities are needed as global warming is changing our climate quickly. Deforestation is one of the major factors speeding up climate change, leading to warmer summers and rising sea levels. It is estimated that nearly 15 billion trees are cut down every year. It has been observed that a group of two people can plant anywhere from 500 to 1,500 seeds per day, depending on certain site conditions like terrain complexity. However, using a mechanical planter, a group of up to 3 people can plant 4,000 to 6,000 seeds per day [3]. UAVs can accelerate this process, as discussed in Ref. [34] by Flash Forest, a Canadian-based company. They claim that their UAV can plant 10,000 to 20,000 seed pods per day. Similarly, another emerging startup called Lord of Trees claims that their UAV can plant up to 1,60,000 seed pods per day [35]. This demonstrates how effective UAVs can be in tree plantation. UAVs can fly over an area, assess it through an onboard image processing system, and release seeds if they detect an area suitable for planting trees, as depicted in Figure 6.10. They can also monitor forest cover and detect any abnormalities. Yuan, Liu, & Zhang [36] discusses how UAVs can be used in forest fire detection and tracking.

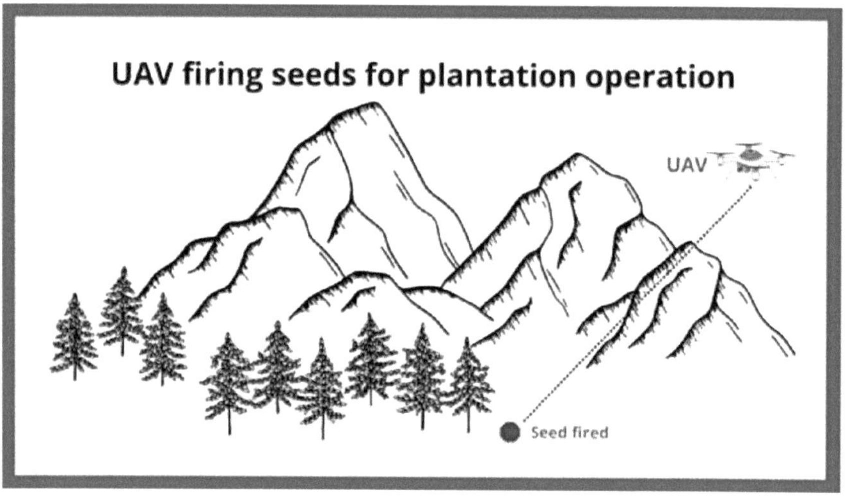

FIGURE 6.10 UAV assisted forestation.

6.8 REINFORCEMENT LEARNING IN UAVS

Regardless of extensive research and development in UAVs and their control systems, autonomous flight is still a challenging task for UAVs. To achieve a high level of autonomy in UAV flying, we need complex mathematical models for UAV computers that can adapt to our dynamic world. Autonomous control systems are also referred to as autopilot controls, which are typically implemented using Proportional-Integral-Derivative (PID) control systems that have proven to be promising in robotics but lag in UAV implementation [37]. To develop a high-level control method that can adapt to a changing environment for UAV implementation, researchers have worked on reinforcement learning-based models. Imanberdiyev, Fu, Kayacan, & Chen [38] discusses the performance of a Reinforcement Learning-based model for autonomous UAV flight, which even outperforms Q-learning-based methods. Reinforcement Learning shows promising results in developing self-learning routing protocols that hold the future of modern civilization.

6.9 FUTURE DIRECTIONS

Literature study shows that very little work has been done in regard to developing self-learning routing protocols for UAV autonomous flight. The TEXPLORE model, as explained in Ref. [38], is one of the best examples of a reinforcement learning-based self-learning model. However, there are many improvements that can be made in autonomous navigation and operation algorithms for UAVs (Figure 6.11).

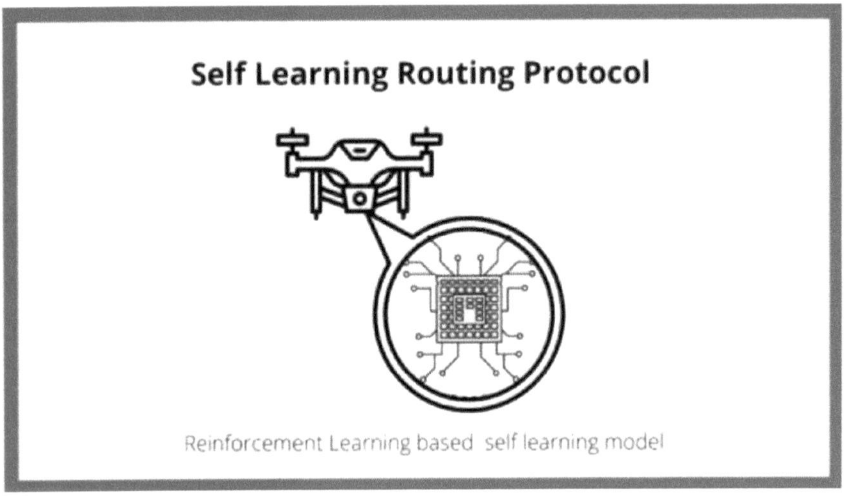

FIGURE 6.11 Reinforcement learning for self-learning routing protocol.

6.10 CONCLUSION

From the study above, it can be concluded that UAVs have numerous applications in our future. They will be very closely used in our daily life. However, this extensive use of UAVs for several critical applications like emergency medical assistance, disaster management, search and rescue missions, package delivery, etc., cannot be materialize until we have made significant advancements in self-learning routing protocols. UAV autonomy is key to all these applications. There has been some work done towards the development of self-learning UAV algorithms [38–42]. However, there is still room for significant improvements in this domain which can help tap the full potential of UAV-assisted autonomous operations in the future.

KEYWORDS

- **disaster management**
- **emergency medical assistance**
- **flying ad-hoc networks**
- **package delivery**
- **proportional-integral-derivative**
- **routing protocols**
- **self-learning routing protocols**
- **unmanned aerial vehicles**

REFERENCES

1. Khan, I. U., Asrin, A., Abdul, J., Bisma, B., Muhammad, A. A., & Fazal, S., (2022). A novel design of FANET routing protocol aided 5G communication using IoT. *Journal of Mobile Multimedia*, 1333–1354.
2. Khan, I. U., Abul, H. M., Muhammad, F., Jeonghwan, G., & Muhammad, A. A., (2022). Improved sequencing heuristic DSDV protocol using nomadic mobility model for FANETS. *Comput., Mater. Continua., 70*(2), 3653–3666.
3. Khan, I. U., Syeda, Z. N. Z., Asrin, A., Shahbaz, A. I., Ijaz, M. Q., Muhammad, A. A., & Syed, B. H. S., (2020). Reinforce based optimization in wireless communication technologies and routing techniques using internet of flying vehicles. In: *The 4th International Conference on Future Networks and Distributed Systems (ICFNDS)* (pp. 1–6).
4. Khan, I. U., Qureshi, I. M., Aziz, M. A., Cheema, T. A., & Shah, S. B. H., (2020). Smart IoT control-based nature inspired energy efficient routing protocol for flying ad hoc network (FANET). In: *IEEE Access* (Vol. 8, pp. 56371–56378). doi: 10.1109/ACCESS.2020.2981531.
5. Hassan, M. A., Syed, I. U., Inam, U. K., Syed, B. H. S., Abdus, S., & Arbab, W. U. K., (2020). Unmanned aerial vehicles routing formation using fisheye state routing for flying ad-hoc networks. In: *The 4th International Conference on Future Networks and Distributed Systems (ICFNDS)* (pp. 1–7).
6. Begum, S., Yao, N., Syed, B. H. S., Inam, U. K., & Satish, A., (2019). Research challenges in the internet of things (IoTs). In: *Recent Trends and Advances in Wireless and IoT-enabled Networks* (pp. 263–274). Springer, Cham.
7. Khan, I. U., Syed, B. H. S., Lei, W., Muhammad, A. A., Thompson, S., & Neeraj, K., (2021). Routing protocols & unmanned aerial vehicles autonomous localization in flying networks. *International Journal of Communication Systems*, e4885.

8. Khan, I. U., Ryan, A., Hasan, J. A., Mohammed, A. I., Muhammad, A. A., Vinh, T. H., & Tanweer, A. C., (2021). RSSI-controlled long-range communication in secured IoT-enabled unmanned aerial vehicles. *Mobile Information Systems, 2021.*
9. Mohamed, N., Al-Jaroodi, J., Jawhar, I., Idries, A., & Mohammed, F., (2020). Unmanned aerial vehicles applications in future smart cities. *Technological Forecasting and Social Change, 153*, 119293.
10. Perkins, C. E., & Bhagwat, P., (1994). Highly dynamic destination-sequenced distance-vector routing (DSDV) for mobile computers. *ACM SIGCOMM Computer Communication Review, 24*(4), 234–244.
11. Jacquet, P., Muhlethaler, P., Clausen, T., Laouiti, A., Qayyum, A., & Viennot, L., (2001). Optimized link state routing protocol for ad hoc networks. In: *Proceedings; IEEE International Multi Topic Conference, 2001: IEEE INMIC 2001; Technology for the 21st Century* (pp. 62–68). IEEE.
12. Perkins, C., Belding-Royer, E., & Das, S., (2003). *RFC3561:* RFC 3561: Ad hoc on-demand distance vector (AODV) routing, Experimental.
13. Johnson, D. B., & Maltz, D. A. (1996). Dynamic Source Routing in Ad Hoc Wireless Networks. In: Imielinski, T., & Korth, H. F. (eds). *Mobile Computing. The Kluwer International Series in Engineering and Computer Science* (vol. 353, pp. 151–183). Springer, Boston, MA.
14. Park, V. D., & Corson, M. S., (1997). A highly adaptive distributed routing algorithm for mobile wireless networks. In: *Proceedings of INFOCOM'97* (Vol. 3, pp. 1405–1413). IEEE.
15. Malhotra, A., & Kaur, S., (2022). A comprehensive review on recent advancements in routing protocols for flying ad hoc networks. *Transactions on Emerging Telecommunications Technologies, 33*(3), e3688.
16. Iordanakis, M., Yannis, D., Karras, K., Bogdos, G., Dilintas, G., Amirfeiz, M., Colangelo, G., & Baiotti, S., (2006). Ad-hoc routing protocol for aeronautical mobile ad-hoc networks. In: *Fifth International Symposium on Communication Systems, Networks and Digital Signal Processing (CSNDSP)* (pp. 1–5). Cite seer.
17. Sahingoz, O., (2013). Mobile networking with UAVs: Opportunities and challenges. In: *Unmanned Aircraft Systems (ICUAS), 2013 International Conference* (pp. 933–941).
18. Pederi, Y. A., & Cheporniuk, H. S., (2015). Unmanned aerial vehicles and new technological methods of monitoring and crop protection in precision agriculture. In: *2015 IEEE International Conference Actual Problems of Unmanned Aerial Vehicles Developments (APUAVD)* (pp. 298–301). IEEE.
19. Camp, T., Boleng, J., & Davies, V., (2002). A survey of mobility models for ad hoc network research. *Wireless Communications and Mobile Computing, 2*(5), 483–502.
20. Cheng, X., Dong, C., Dai, H., & Chen, G., (2018). MOOC: A mobility control based clustering scheme for area coverage in FANETs. In: *2018 IEEE 19th International Symposium on" A World of Wireless, Mobile and Multimedia Networks" (WoWMoM)* (pp. 14–22). IEEE.
21. Leonov, A. V., & Litvinov, G. A., (2018). About applying AODV and OLSR routing protocols to relaying network scenario in FANET with mini-UAVs. In: *2018 XIV International Scientific-Technical Conference on Actual Problems of Electronics Instrument Engineering (APEIE)* (pp. 220–228). IEEE.

22. Oubbati, O. S., Atiquzzaman, M., Lorenz, P., Tareque, M. H., & Hossain, M. S., (2019). Routing in flying ad hoc networks: Survey, constraints, and future challenge perspectives. *IEEE Access*, *7*, 81057–81105.
23. Haas, Z. J., & Pearlman, M. R., (2001). Zone routing protocol (ZRP) a hybrid framework for routing in ad hoc networks; chapter 7. *Ad hoc Networking* (2nd edn., Vol. 1).
24. Abdellah, E., Benalla, S. A. I. D., Hssane, A. B., & Hasnaoui, M. L., (2010). Advanced low energy adaptive clustering hierarchy. *IJCSE) International Journal on Computer Science and Engineering*, *2*(07), 2491–2497.
25. Caleffi, M., & Paura, L., (2011). M-DART: Multi-path dynamic address routing. *Wireless Communications and Mobile Computing*, *11*(3), 392–409.
26. Jacquet, P., Muhlethaler, P., Clausen, T., Laouiti, A., Qayyum, A., & Viennot, L., (2001). Optimized link state routing protocol for ad hoc networks. In: *Proceedings. IEEE International Multi Topic Conference, 2001: IEEE INMIC 2001; Technology for the 21st Century* (pp. 62–68). IEEE.
27. Clausen, T., & Jacquet, P., (2003). RFC3626: Optimized Link State Routing Protocol (OLSR), Experimental.
28. Mahesh, B., (2020). Machine learning algorithms: a review. *International Journal of Science and Research (IJSR). [Internet]*, *9*, 381–386.
29. Bonaccorso, G., (2017). *Machine Learning Algorithms*. Packt Publishing Ltd.
30. Bujari, A., Palazzi, C. E., & Ronzani, D., (2017). FANET application scenarios and mobility models. In: *Proceedings of the 3rd Workshop on Micro Aerial Vehicle Networks, Systems, and Applications* (pp. 43–46).
31. Yeong, S. P., King, L. M., & Dol, S. S., (2015). A review on marine search and rescue operations using unmanned aerial vehicles. *International Journal of Marine and Environmental Sciences*, *9*(2), 396–399.
32. Erdelj, M., Natalizio, E., Chowdhury, K. R., & Akyildiz, I. F., (2017). Help from the sky: Leveraging UAVs for disaster management. *IEEE Pervasive Computing*, *16*(1), 24–32.
33. Stelzer, H., (2019). *How to Plant Forest Tree Seedlings.*
34. Mohan, M., Richardson, G., Gopan, G., Aghai, M. M., Bajaj, S., Galgamuwa, G. P., Vastaranta, M., et al., (2021). UAV-supported forest regeneration: Current trends, challenges and implications. *Remote Sensing*, *13*(13), 2596.
35. Nordquist, C., (2021). *Drones the Latest Tool in Reforestation Efforts Following Wildfires.*
36. Yuan, C., Liu, Z., & Zhang, Y., (2015). UAV-based forest fire detection and tracking using image processing techniques. In: *2015 International Conference on Unmanned Aircraft Systems (ICUAS)* (pp. 639–643). IEEE.
37. Koch, W., Mancuso, R., West, R., & Bestavros, A., (2019). Reinforcement learning for UAV attitude control. *ACM Transactions on Cyber-Physical Systems*, *3*(2), 1–21.
38. Imanberdiyev, N., Fu, C., Kayacan, E., & Chen, I. M., (2016). Autonomous navigation of UAV by using real-time model-based reinforcement learning. In: *2016 14th International Conference on Control, Automation, Robotics and Vision (ICARCV)* (pp. 1–6). IEEE.
39. Pham, H. X., La, H. M., Feil-Seifer, D., & Nguyen, L. V., (2018). *Autonomous UAV Navigation Using Reinforcement Learning.* arXiv preprint arXiv:1801.05086.

40. Liu, X., Liu, Y., & Chen, Y., (2019). Reinforcement learning in multiple-UAV networks: Deployment and movement design. *IEEE Transactions on Vehicular Technology*, *68*(8), 8036–8049.
41. Liu, C. H., Chen, Z., Tang, J., Xu, J., & Piao, C., (2018). Energy-efficient UAV control for effective and fair communication coverage: A deep reinforcement learning approach. *IEEE Journal on Selected Areas in Communications*, *36*(9), 2059–2070.
42. Wang, T., Qin, R., Chen, Y., Snoussi, H., & Choi, C., (2019). A reinforcement learning approach for UAV target searching and tracking. *Multimedia Tools and Applications*, *78*(4), 4347–4364.

CHAPTER 7

Smart Intrusion Detection-Enabled FANETs in Healthcare: A Survey

MARIA NAWAZ CHOHAN,[1] SANA NAWAZ CHOHAN,[2] INAM ULLAH KHAN,[3] MUHAMMAD YASEEN AYUB,[4] and ALI HAIDER[4]

[1]*National Defense University, Islamabad, Pakistan*

[2]*Foundation University Institute of Rehabilitation Sciences, Islamabad, Pakistan*

[3]*Department of Electronic Engineering, School of Engineering and Applied Sciences (SEAS), Isra University, Islamabad Campus, Pakistan*

[4]*Department of Computer Science, COMSATS University Islamabad, Attock Campus, Pakistan*

ABSTRACT

Internet of Medical Things is a new concept in the healthcare industry which has merged IoT and medical together. The smart world is embracing modern technologies where e-healthcare is introduced. Most importantly, flying ad hoc networks are utilized in healthcare to speed up communication and transportation. FANET is the combination of unmanned aerial vehicles which have many civil and military applications. Maintaining health records of patients, IoT plays an important role in providing easy access to doctors and patients. However, the healthcare industry has contemporary technologies but lacks sufficient security. Therefore, an

Advances in Autonomous Navigation Through Intelligent Technologies.
Niharika Singh, Thipendra P. Singh, Brian Azzopardi (Eds.)

intrusion detection system will identify cybersecurity attacks that can impact performance metrics within the healthcare industry. This research chapter is based on a comprehensive survey regarding attack detection mechanisms in flying ad hoc networks utilized in the healthcare industry.

7.1 INTRODUCTION

Technology has always amazed mankind in all walks of life. The journey of technology started back then from the first industrial revolution which used water and steam to mechanize production. It is a systematic shift in technology which aims to improve and facilitate mankind in every respect [1]. Currently, we are living in the Fourth Industrial Revolution which is blurring the lines between the physical, digital, and biological spheres by the digitalization in manufacturing, automation, and data exchange. It is the convergence of the real and the virtual world. There are a lot of advantages of this era as it has eased human beings in every regard by providing improved quality of life along with ease of living. This revolution is the most dehumanized form and has the potential to robotize humanity. This revolution has introduced robots and software that have the capabilities to carry out the work that humans used to do. Due to such technological advancements, automation is taking a toll on all aspects of life. From wired to wireless technology, the dependency on such innovation has increased by numbers. The demand for portable and flexible advancements has also increased considerably. Wireless communication between two devices without having any central infrastructure is gaining popularity these days [2]. Wireless ad-hoc networks, particularly used in monitoring and surveillance applications, are worth mentioning. Flying ad-hoc networks (FANETs) are used to govern the autonomous movement of unmanned aerial vehicles (UAVs). Multiple UAVs are connected together, and an ad-hoc system is established between them. Such communication between UAVs negates the need for the installation of complex hardware among the UAVs. These ad-hoc systems remain functional even if one of the UAVs' communication links breaks down. In the field of medicine, such technological advancements are required for treatment, assistance in healthcare procedures, and maintaining datasets. These advancements in healthcare aim to provide cost-effective and efficient solutions to patients. The acceptance of such wireless technologies by healthcare industries enables a more mobile approach to healthcare information systems [3]. Despite the advantageous nature of these wireless systems, the limited security has raised concerns. For this reason, an intrusion detection system (IDS) was introduced. The sole purpose of this system is to ensure a reliable and secure network system. This system detects any suspicious activity that might compromise the patient's vital information. Figure 7.1 shows the FANET topological structure.

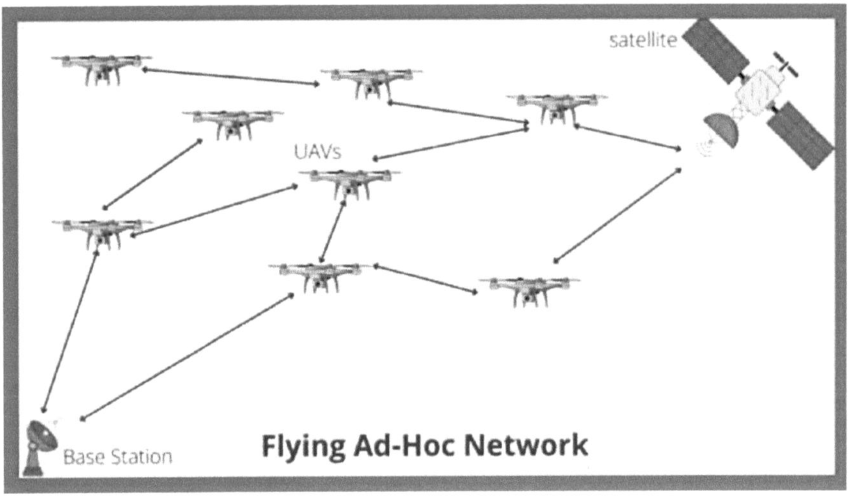

FIGURE 7.1 FANET topological structure.

7.2 LITERATURE REVIEW

In the last two decades, there has been a lot of research in the domain of FANETs and their applications. In the year 2012, the US military increased their research and production budget for UAVs to $4.2 billion [20]. The concept of smart cities, which is becoming a reality, also heavily incorporates UAV technology for different types of services such as traffic monitoring, surveillance, and agricultural monitoring. When UAVs are used in such close proximity to humans, there are certain risks involved if the UAV is compromised, especially when the UAV is being used for healthcare purposes. FANETs are best suited for limited-time rescue operations or situations where medical aid needs to be provided quickly. UAVs can prove to be fast and efficient in such operations, as depicted in Figure 7.2.

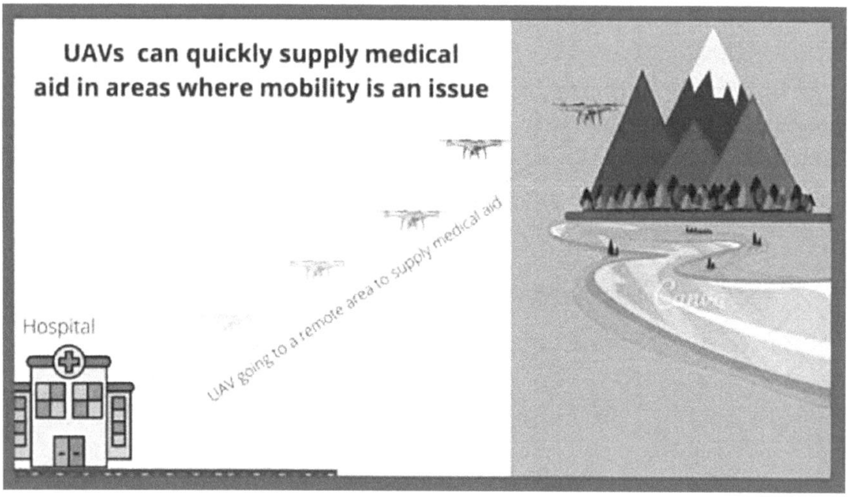

FIGURE 7.2 UAVs can be efficient in delivering medical aid in remote areas.

Thus, if a node (UAV) is compromised, its consequences can be fatal. Hence, a strong security mechanism is needed. This section gives a brief background on the work that has already been done in this regard. With the sudden surge in COVID-19 cases in 2020, the world has witnessed a tremendous increase in UAV and IoT-based applications in general. The Chinese government has deployed drones for several purposes such as crowd monitoring, vaccine supply to remote areas, medicine delivery, etc. This reduced the burden on hospitals [15]. For such critical operations, it is essential to secure the nodes (UAVs) from any type of hacking or hijacking attempts. FANETs have a unique design architecture which makes it difficult to establish a strong defense against dynamically changing security threats. Researchers [16] have discussed how the integration of drones into daily life can pose severe security and privacy risks [17]. In FANETs, drones heavily rely on their autonomous flying features which are integrated into their onboard circuit, thus minimizing human intervention. Researchers worry that such systems are prone to cyber-attacks [18, 19]. There can be several different cyber-attacks that can affect the performance of FANETs as discussed in Refs. [21, 22].

7.3 INTRUSION DETECTION SYSTEM FOR FANETS

An intrusion detection system (IDS) is a system installed that not only monitors but also keeps check and balance on network traffic. The sole purpose of this system is to detect and clear out any suspicious activity. After detection of any doubtful activity, the system alerts and notifies the desired authorities. It acts as an invigilating body that scans every system for any malicious activity or information in the network. All of this malicious information is reported to an administrator to carry out the required action. It could also be gathered centrally by using a system known as Security Information and Event Management (SIEM). This system helps in differentiating between suspicious and doubtful activity from false alarms. By using the SIEM system, one can collect outputs from multiple sources and filter out the information. It also helps in disposing of any information that is not of any danger. Organizations that install IDS systems set out the criteria that recognize normal traffic on the network as compared to malicious activity. Organizations need to filter out their IDS products at the time of installation. This system warns of any malicious activity detected by sending notifications. Figure 7.3 illustrates the intrusion detection system for UAV networks.

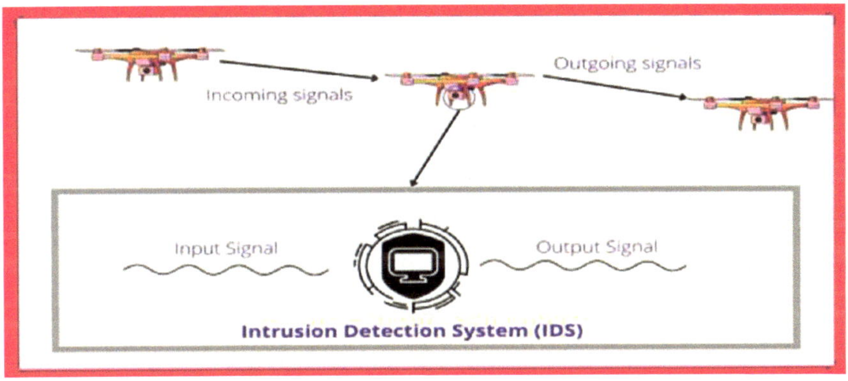

FIGURE 7.3 IDS for UAV-network.

7.4 TYPES OF INTRUSION DETECTION SYSTEMS

There are few types of intrusion detection system:

1. **Network-based IDS (NIDS):** It is used to examine traffic from all devices on the network. It executes an observation of every passing information on the entire network and then contrasts the information with the collection of known attacks. An alert is sent instantly to the administrator if the system detects any harmful or

abnormal behavior within the subnet. A good example of NIDS is by installing it on the system containing firewalls to see if a spy is trying to crack the firewall.

2. **Host-based IDS (HIDS):** It works on independent hosts or devices on the network. It basically keeps track of all incoming and outgoing packets from the device as it takes a snapshot of existing system files and compares it with the previous snapshot [12]. If someone tries to edit or delete the analytical system files or if it detects any malicious activity, it immediately alerts the administrator to investigate. An example of HIDS usage can be observed on mission-critical machines which may or may not change their layout.
3. **Protocol-based Intrusion Detection System (PIDS):** It consists of a system or agent that is present at the front end of the server, which works to control and interpret the protocol between the user and the server [12]. It helps in securing the web server by regularly observing and accepting the related HTTP protocol from the HTTPS protocol stream.
4. **Application Protocol-based Intrusion Detection System (APIDS):** It is a system that is present within a group of servers. It monitors and interprets the communication on application-specific protocols and recognizes intrusions. For example, it would monitor the SQL protocol specific to the middleware as it interacts with the database in the web server.
5. **Hybrid Intrusion Detection System:** It develops a complete view of the network system by combining host agent or system data with network information. It is the most effective among intrusion detection systems. Prelude is an example of a Hybrid intrusion detection system.

7.5 EXISTING IT SOLUTIONS TO HEALTHCARE

The advantages of technology are undeniable. Technology has eased the paths for humans in all fields. IT has facilitated not only in the domains of warfare but has also played a considerable role in the field of science.

It has a major impact on the dissemination of medical information and knowledge to healthcare practitioners [3]. It has helped healthcare professionals to search for any useful information available in online formats. These online services allow paramedic staff to have quick access to reliable information. This also helps healthcare providers to avoid any useless information on these platforms and to differentiate between new authentic information and outdated and less reliable data. Doctors and healthcare providers can not only share this information with others but can also keep a record of it [5]. It can also be used to send complex medical files across the internet. These electronic healthcare initiatives can help patients to have online access to their medical records. Patients are also informed about their health and ways to manage the disease. These initiatives have played a remarkable role in disease prevention and rehabilitation [1]. With the help of e-health, patients can seek help about their signs and symptoms and treatment options available. Patients can gain knowledge about the disease and ways to prevent disease development by getting proper information from specialized doctors. Some of the major IT technologies that have been introduced in the healthcare industry are discussed in subsections.

7.5.1 TELEMEDICINE

The main idea behind this initiative is to provide readily available and best treatment based on clinical expertise and knowledge. It allows patients to get treatment in accordance with their perspective and preferences. These websites are readily available on TV, mobile phones, and wireless devices to interact with patients in their home-based capacity [4]. Telemedicine services help patients to schedule their appointments online. This allows the surgeon to perform remote surgical procedures directed by a surgeon. This happens through high-bandwidth technologies and video cameras. The most important element of telemedicine is that it has reached and helped out patients residing in remote areas [1]. With the use of web TV and cameras, doctors can take a proper history from patients and diagnose patients based on investigations and signs and symptoms. Patients can get the desired treatment plan without having to leave their homes. Telehealth has helped elderly patients, patients with chronic diseases, and handicapped patients through online check-ups.

7.5.2 DECISION SUPPORT SYSTEMS (DSS)

This system is used to save the gold standard diagnostic techniques used to manage a disease. This also enables the practitioner to keep up the record of the patient and to cross-check with the previous records available [4]. With the availability of the previous records, doctors can estimate the improvement in disease and disease progression. This can improve the patient's health by sending reminders to the patient regarding the medicine dosage and certain preventive measures to avoid disease progression [5]. This can be used in conjunction with telehealth to communicate regarding disease and patient education and care.

7.5.3 INFORMATION WAREHOUSING AND DATA MINING TECHNIQUES

Pharmaceutical companies use information warehousing for marketing purposes. Healthcare providers are using this technique for diagnosing a disease and treatment of the patient. Data mining technique is more cost-effective and efficient. This enables healthcare professionals to provide quality care at a lower cost to patients living in remote areas [4]. The use of these two techniques is increasing day by day.

7.6 CYBER-ATTACKS USING UAVS

Possibility of cyber-attacks against medical BAN and implantable medical devices is increasing with every passing day. Due to their small size and easy availability over secured terrain, these UAVs are an ideal platform for wireless and IoT cyber-attacks [7]. Because of their high precision, UAVs are posing threats for the cyber-attacks as they can invade physical security controls and can hijack medical data and devices available. Such systems can work undercover to penetrate the desired territory and to perform any attack with high precision. Because of their small size and hardly detectable nature, these UAVs can glide over a hospital and clinic to conduct any sort of cyber-attack. The common attacks conducted by the use of UAVs against healthcare devices are as in subsections.

7.6.1 DE AUTHENTICATION ATTACK

This is a form of denial of service (DoS) attack in which an intruder intentionally knackers the site's resources leading to the denial of legitimate users' access [13]. First, the attacker sends a series of unauthenticated frames to the clients and requests them to disconnect from the access point. Secondly, the same attacker sends a series of unauthenticated packets to AP to reauthenticate all the connected clients. This re-authentication between client and AP launches a cyber-attack that disconnects all the clients. Figure 7.4 shows denial of service attack or third-party attack in UAV-network.

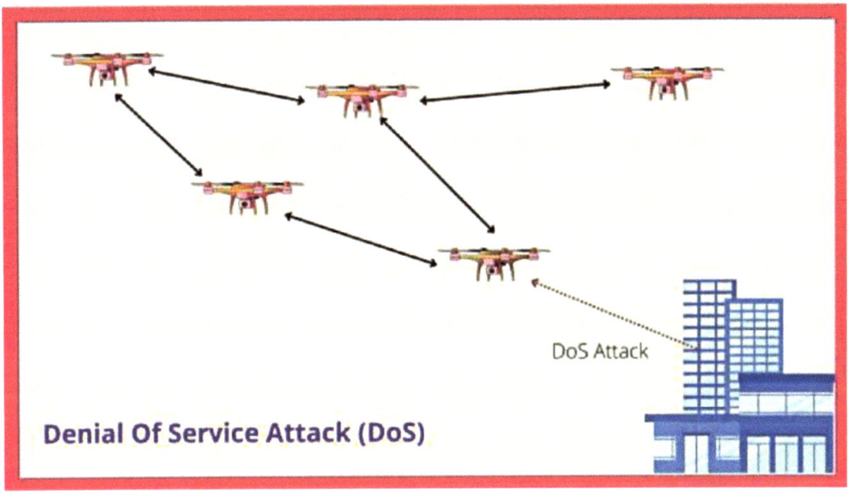

FIGURE 7.4 DoS attack in UAV-network.

7.6.2 *STEPPINGSTONE ATTACK*

This type of attack uses multiple UAVs to launch an attack. These UAVs are connected to each other via a simple mobile hotspot. The UAVs receive attack commands from an attacker who is independently controlling all these devices. The attack command is then circulated among these aerial vehicles [13]. Figure 7.5 shows the attack scenario in a UAV network.

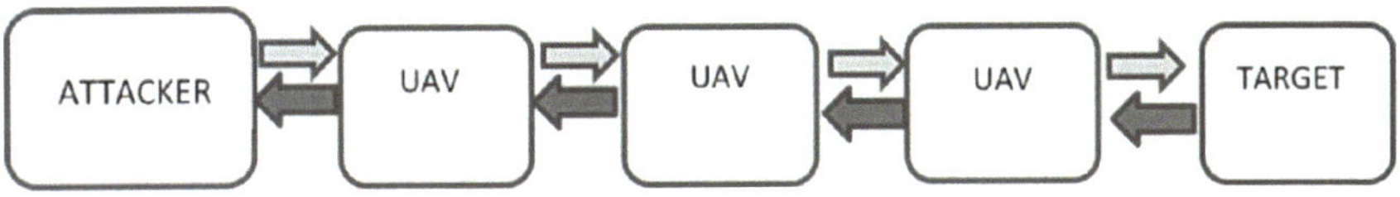

FIGURE 7.5 Attacker or intruder scenario in UAV-network.

7.6.3 DRONE-IN-THE-MIDDLE (DITM) ATTACK

In this case drone can be utilized as third party, therefore commonly called drone in the middle attack.

7.6.4 CLOUD ASSISTED UAV ATTACK

UAVs are used by the attacker to store the data that is hacked. This allows the attacker to have access to all the data at any time and any place. UAVs with limited storage cannot perform the complex computation to retrieve all the important information [13]. For these reasons, cloud-assisted UAVs are used to move the data easily to the cloud with minimal battery usage. Cloud-assisted UAVs are more preferred over simple UAVs when storage burden and limited battery storage are the problems.

7.6.5 EVIL TWIN ATTACK

This works on the same principle as a DitM attack. It takes over as the receiver for BAN and IMD. Firstly, the attacker de-authenticates the clients connected to the AP through de-authentication probes [13]. Secondly, the attacker masquerades as a legitimate AP and launches a fake access point.

7.6.6 WI-FI-PHISHING

This attack uses the masquerading technique and is used on Wi-Fi networks. Such attacks are used to hijack login passwords and medical account information. This shows up as a fake login page which forcibly allows clients to enter all the information.

7.7 FUTURE CHALLENGES

- To make FANET a more reliable and efficient tool, proper coordination between these unmanned aerial vehicles should be ensured. Proper updates should be done on a regular basis. New algorithms and methods should be introduced to properly organize clusters of UAVs.

- To improve the quality of service, proper provisions regarding delay, packet loss, and bandwidth need to be possible [9].
- Improving coordination between UAVs and manned aerial vehicles should be ensured.
- Ensure a 5G network system for better communication between these vehicles [2].
- High dynamic network topology in FANETs as a result of high mobility causes low-quality links [10]. Certain disconnections among these will cause poor-quality routing performance [7, 8].
- Reliability of routes also poses a real challenge because of the very high mobility in FANET [11].
- High cost, regular updates, and low residual energy are the major concerns of these FANETs.
- Education of paramedical staff regarding the use of Intrusion Detection Systems (IDS) to avoid any future cyber-attacks. Proper education regarding the threats and risks related to data loss should be done.
- Proper education regarding the types and severity of attacks should be given.
- Dependency on technology is the major challenge faced by healthcare professionals, i.e., mobile robotic surgery and remote diagnosis. The major drawback of such a system is that it will only diagnose and analyze up to the capacity of information installed.
- This era is the most dehumanized form and has the potential to robotize humanity. This revolution has introduced robots and software that have the capabilities to carry out the work that humans used to do. For example, an app like Human DX is for online self-diagnosis. This app is replacing doctors and hospitals.
- With the digitalization of work, there is an increased security risk from hackers. Malware, password hacking, and cross-site attacks are the major threats.
- These risks have increased with digitalization because anyone can access our data without our consent by hacking it and then use that data for their personal gain. We never know if our data has been hacked or is being viewed.

7.8 CONCLUSION

From the first revolution to the fourth industrial revolution, technology has come a long way. From wired to wireless technology, the dependency on such innovation has increased exponentially [6]. Multiple UAVs are connected together, and an ad-hoc system is established between them. The acceptability of such wireless technologies by healthcare industries enables a more mobile approach to healthcare information systems. Regardless of the benefits of such wireless systems, limited security is still one of the major concerns. Cyber-attacks using UAVs are major threats to hospital-based vital information. For these reasons, the intrusion detection system (IDS) was introduced. The main goal of such a system is to guarantee a reliable and secure network system. This system identifies any suspicious activity that could potentially jeopardize the patient's vital information. Telemedicine and decision support systems (DSS) are the biggest innovations in the field of science. Such initiatives have helped healthcare professionals to have quick access to reliable and vital information. These initiatives have played a remarkable role in disease detection and prevention. Future challenges faced by such advanced systems must be kept under consideration. Emphasis should be placed on educating paramedical staff regarding the use of intrusion detection systems. Proper coordination among UAVs, improving quality of service, reliability, and low residual energy are some of the major concerns of these FANETs. We need advanced solutions for such advanced technology.

KEYWORDS

- **advanced technology**
- **cyber-attacks**
- **decision support systems**
- **e-healthcare**
- **flying ad-hoc networks**
- **intrusion detection systems**
- **unmanned aerial vehicles**

REFERENCES

1. Khan, I. U., Muhammad, A. H., Mohammad, D. A., Mohammed, A. I., Hasan, J. A., Ryan, A., & Vinh, T. H., (2021). Monitoring system-based flying IoT in public health and sports using ant-enabled energy-aware routing. *Journal of Healthcare Engineering, 2021*.
2. Ullah, H., Nithya, G. N., Adrian, M., Chris, N., Paul, M., & Maria, C., (2019). 5G communication: An overview of vehicle-to-everything, drones, and healthcare use-cases. *IEEE Access, 7*, 37251–37268.
3. Iwendi, C., Joseph, H. A., Cresantus, B., & Desire, N., (2021). Security of things intrusion detection system for smart healthcare. *Electronics, 10*(12), 1375.
4. Srivastava, P. K., & Sandeep, S., (2004). A cost-effective solution for telemedicine in rural health care using mobile ad hoc networks. In: *E-Tech 2004* (pp. 109–113). IEEE.
5. Mukhopadhyay, A., & Dipan, G., (2020). FANET based emergency healthcare data dissemination. In: *2020 Second International Conference on Inventive Research in Computing Applications (ICIRCA)* (pp. 170–175). IEEE.
6. Azevedo, M. I. B., Carlos, C., Eylon, M. T., Tassio, C. C., & José, J., (2020). Wireless communications challenges to flying ad hoc networks (FANET). *Mobile Computing*, 3.
7. Khan, M. A., Alamgir, S., Ijaz, M. Q., & Inam, U. K., (2017). Flying ad-hoc networks (FANETs): A review of communication architectures, and routing protocols. In: *2017 First International Conference on Latest Trends in Electrical Engineering and Computing Technologies (INTELLECT)* (pp. 1–9). IEEE.
8. Bekmezci, I., Ozgur, K. S., & Şamil, T., (2013). Flying ad-hoc networks (FANETs): A survey. *Ad Hoc Networks, 11*(3), 1254–1270.
9. Nadeem, A., Alghamdi, T., Yawar, A., Mehmood, A., & Siddiqui, M. S., (2018). A review and classification of flying Ad-Hoc network (FANET) routing strategies. *Journal of Basic and Applied Scientific Research, 8*(3), 1–8.
10. Zafar, W., & Bilal, M. K., (2016). Flying ad-hoc networks: Technological and social implications. *IEEE Technology and Society Magazine, 35*(2), 67–74.
11. Boukerche, A., Turgut, B., Aydin, N., Ahmad, M. Z., Bölöni, L., & Turgut, D. (2011). Routing protocols in ad hoc networks: A survey. *Computer networks*, *55*(13), 3032–3080.
12. Bace, R. G., & Peter, M., (2001). *Intrusion Detection Systems,* 201.
13. Sethuraman, S. C., Vaidehi, V., & Steven, W., (2020). Cyber-attacks on healthcare devices using unmanned aerial vehicles. *Journal of Medical Systems, 44*(1), 1–10.
14. Khan, M. F., Kok-Lim, A. Y., Rafidah, M. N., & Muhammad, A. I., (2020). Routing schemes in FANETs: A survey. *Sensors, 20*(1), 38. https://doi.org/10.3390/s20010038.
15. Kumar, S., Rathore, N. K., Prajapati, M., & Sharma, S. K. (2023). SF-GoeR: an emergency information dissemination routing in flying Ad-hoc network to support healthcare monitoring. *Journal of Ambient Intelligence and Humanized Computing*, *14*(7), 9343–9353.
16. Altawy, R., & Youssef, A. M., (2016). Security, privacy, and safety aspects of civilian drones: A survey. *ACM Transactions on Cyber-Physical Systems*, *1*(2), 1–25.

17. Rani, C., Modares, H., Sriram, R., Mikulski, D., & Lewis, F. L., (2016). Security of unmanned aerial vehicle systems against cyber-physical attacks. *The Journal of Defense Modeling and Simulation, 13*(3), 331–342.
18. Kim, A., Wampler, B., Goppert, J., Hwang, I., & Aldridge, H., (2012). Cyber attack vulnerabilities analysis for unmanned aerial vehicles. In: *Infotech@ Aerospace 2012* (p. 2438).
19. Krishna, C. L., & Murphy, R. R. (2017, October). A review on cybersecurity vulnerabilities for unmanned aerial vehicles. In *2017 IEEE international symposium on safety, security and rescue robotics (SSRR)* (pp. 194–199). IEEE.
20. Hartmann, K., & Steup, C., (2013). The vulnerability of UAVs to cyber-attacks – An approach to the risk assessment. In: *2013 5th International Conference on Cyber Conflict (CYCON 2013)* (pp. 1–23). IEEE.
21. Hussain, A., Heidemann, J., & Papadopoulos, C., (2003). A framework for classifying denial of service attacks. In: *Proceedings of the 2003 Conference on Applications, Technologies, Architectures, and Protocols for Computer Communications* (pp. 99–110).
22. Lakhina, A., Crovella, M., & Diot, C., (2004). Diagnosing network-wide traffic anomalies. *ACM SIGCOMM Computer Communication Review, 34*(4), 219–230.

CHAPTER 8

Concept and Algorithms of Mobile Devices

SAVITA,[1] DEEPTI SINGH,[2] and ANKITA VERMA[2]

[1]*Gurukul Kangri (Deemed to be University), Haridwar, Uttarakhand, India*

[2]*Department of CSE&IT, Jaypee Institute of Information Technology, Noida, Uttar Pradesh, India*

ABSTRACT

Mobile devices, such as smartphones and tablets, use a combination of hardware and software to perform a variety of tasks. The hardware includes components such as a processor, memory, storage, and various sensors. The software includes the operating system, as well as various applications that run on the device. Studying the concepts and algorithms of mobile devices can be beneficial for those who want to develop, design, research, or use mobile technology in their field. Understanding the algorithms used in mobile devices can help developers create more efficient and effective apps that run well on these devices. Hardware engineers optimize the use of resources such as battery and memory, as well as understanding how to design new sensors and capabilities. The concepts and algorithms used in mobile devices are applied in a wide range of applications, including communications, location-based services, power management, security, navigation system, entertainment, and many more. Mobile devices use a wide variety of algorithms to enable their many features and capabilities such as location-based algorithms to determine the device's location

Advances in Autonomous Navigation Through Intelligent Technologies.
Niharika Singh, Thipendra P. Singh, Brian Azzopardi (Eds.)

using GPS, Wi-Fi, or cellular network data, communication algorithms to transfer data between the device and cellular or Wi-Fi networks, multitasking algorithms to manage the running of multiple apps simultaneously, power management algorithms to optimize the use of the device's battery, security algorithms to secure the device and protect the user's data, such as encryption algorithms and biometric authentication, machine learning algorithms used in various applications such as image recognition, natural language processing, and speech recognition, and graphical algorithms to render images, video, and 3D graphics on mobile devices.

8.1 INTRODUCTION

The widespread adoption of mobile devices for both personal and professional use has increased the market value of these gadgets significantly. People today utilize a wide variety of mobile devices, each with their own unique design and set of features, all of which are constantly being improved upon. When a person uses a mobile phone to make and receive phone calls, send text messages, and access the Internet, they are engaging in mobile technology. It's the medium by which information is transmitted from one system to another. Mobile technology includes everything from wireless two-way communication systems to portable computers and networking hardware. Cellular communication systems and other similar domains make heavy use of mobile technology for two-way communications, utilizing a network architecture that permits simultaneous data transmission from several transmitters across a shared channel. This technology allows multiple users to share a single frequency by decreasing the possibility of interference from multiple sources. Through time, the channel has undergone a transformation. This is quickly growing; its uses are broadening; and it is gradually displacing other similar communication options on the market, such as postal services and landlines. The evolution of mobile technology has resulted in devices that can do much more than make and receive phone calls, such as provide GPS navigation, surf the web, play games, and communicate with other users in real time via instant messaging apps. Consequently, many believe that wireless networking and mobile computing will determine the course of computer technology in the future.

The utilization of mobile devices such as tablets and small personal computers is booming. In the ensuing years, the smartphone system has evolved into a multifunctional computer capable of GPS navigation, video gaming, web browsing, and instant messaging, among other things. With the introduction of lightweight devices such as tablets, mobile technology has grown more prevalent. The wireless systems that link these mobile devices are also referred to as mobile networks. They make it possible for mobile phones to communicate with one another in terms of sound, text, and (mobile) programs. There are now more than 3 billion smartphone users, and by 2022, the number of mobile workers worldwide is projected to increase to 1.87 billion. Mobile technology refers to any portable electronic device that can connect to the internet. These now include smartphones, tablets, some iPods, and laptops, but that list will grow in the coming years.

8.1.1 WHAT IS A MOBILE DEVICE?

Any portable electronic equipment can be referred to as a "mobile device." These gadgets often fit comfortably in one hand and are made to be carried around easily. The capabilities of modern smartphones, tablets, and e-readers make them viable alternatives to traditional computers in many situations.

Numerous mobile devices exist, each with its own set of features and capabilities. Many mobile devices are depicted in Figure 8.1.

FIGURE 8.1 Mobile devices are available with different shapes and sizes.

8.2 TYPES OF MOBILE DEVICES

A broad category of portable electronic gadgets falls under the umbrella term "mobile device." The term "mobile device" is commonly used to refer to wireless Internet-enabled handheld computers. Some people, however, include conventional MP3 players and internet-enabled digital cameras under the mobile gadgets category. Figure 8.2 depicts some of the cases that fall within this category.

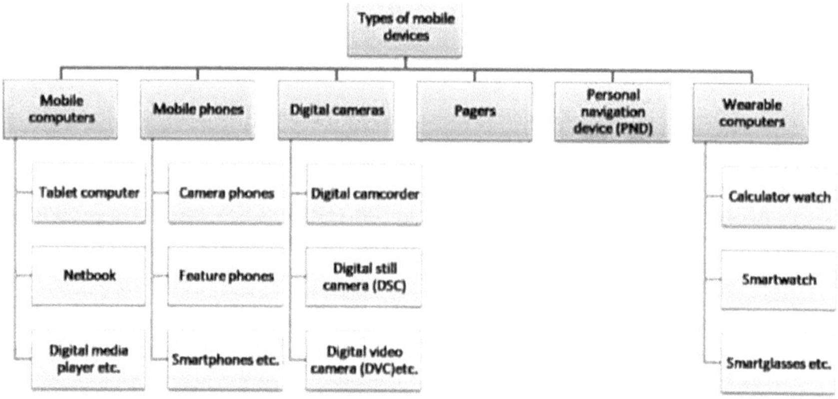

FIGURE 8.2 Types of mobile devices.

8.2.1 MOBILE COMPUTERS

Completely self-contained, mobile computers have all the devices and interfaces required to scan barcodes and manage and store the resulting data. A mobile computer is a portable device that has a full-fledged computer in a compact package. This generally includes a screen, Wi-Fi, an OS (Windows® or AndroidTM), a keyboard (virtual or physical), power, and a high-quality scanning engine. All Datalogic mobile computers have these capabilities and more thanks to their tough enclosures that allow them to survive drops from a variety of heights and continue functioning normally. The gadgets can function in any professional setting because of their durable housing, which is impervious to shocks, waterjets, and dust. Mostly used mobile computers are tablet computer, Netbook, Digital media player, Enterprise digital assistant, Graphing calculator, Handheld game console, Handheld PC, Laptop, Mobile Internet device (MID), Personal digital assistant (PDA), Pocket calculator, Portable media player, Ultra-mobile PC, etc.

8.2.2 MOBILE PHONES

A mobile phone is a portable telephone that allows users to make and receive calls over a radio frequency link while moving within a phone service area. The radio frequency link provides access to the Public Switched Telephone Network (PSTN) by establishing a connection with a mobile phone operator's switching infrastructure. In North America, mobile phones are known as cell phones or cellular telephones because modern mobile telephone services employ a cellular network architecture [1].

1. **Camera Phones:** It is a mobile device with one or more integrated digital cameras that can take pictures and frequently record videos. The final image can also be conveniently and wirelessly transmitted. The Kyocera Visual Phone VP-210, which was made available in Japan in May 1999, was the first commercial phone with a color camera [2]. The charge-coupled device (CCD), developed by Willard S. Boyle and George E. Smith at Bell Labs in 1969 [3] and based on MOS capacitor technology, was the first semiconductor image sensor [4]. Tsutomu Nakamura's team at Olympus later created the NMOS active-pixel sensor in 1985 [5], which prompted Eric Fossum's group at the NASA Jet Propulsion Laboratory to create the CMOS active-pixel sensor (CMOS sensor) in 1993 [6]. Compared to standalone digital cameras, camera phones are typically smaller and easier to use. Point-and-shoot camera sales peaked around 2010 and then started to fall as a result of the continuous rise in camera phone sales during the smartphone era [7]. Compact point-and-shoot cameras are being gradually replaced by smartphones due to the simultaneous advancement of smartphone camera technology and other multipurpose advantages.
2. **Feature Phones:** When referring to mobile phones with fewer functions than a modern smartphone, the term "feature phones" is frequently used. In addition to basic multimedia and Internet capabilities and other services supplied by the user's cellular service provider, feature phones typically provide voice calling and text messaging capabilities. A feature phone is more advanced than a mobile phone with merely voice calls and text messaging capabilities. Typically, the software and user interface of feature phones and basic mobile phones are proprietary and customized.

Smartphones, on the other hand, run a mobile operating system that frequently shares characteristics with other devices [8].

3. **Smartphones:** These have various distinctive qualities. The International Telecommunication Union tallies the number of active mobile broadband subscriptions with an Internet connection (which includes tablets, etc.). In industrialized nations, smartphone usage has now surpassed that of preceding mobile technology. However, they account for approximately 50% of mobile communications in developing countries.

The in-depth assessments of hardware and software characteristics of smartphones and feature phones are presented in Table 8.1.

8.2.3 DIGITAL CAMERAS

A digital camera is a device that captures and stores digital photographs. Film-based cameras have been supplanted by digital cameras as the majority of those manufactured today [9]. Smartphones and other mobile devices typically incorporate digital cameras with capabilities and features comparable to or beyond those of dedicated cameras (which are still available). High-end, high-definition, specialized cameras are still commonly used by professionals and amateurs who aspire to produce better photographs. The most often used digital cameras are the digital camcorder, digital still camera (DSC), digital video camera (DVC), and front-facing camera. A camcorder is a portable video recording device whose components are self-contained. Standard features include a left-mounted articulating screen, a right-mounted belt for better carrying, a user-facing hot-swappable battery and recording media, and an internal, silent optical zoom lens. A video camera is an optical device used to record moving images (as opposed to a movie camera, which records images on film). After having been improved for television production, video cameras quickly found other uses. Typically, cameras, mobile phones, smartphones, tablets, laptops, and even some handheld video game consoles feature a front-facing camera, sometimes known as a selfie camera.

Tablets, smartphones, and other mobile devices typically contain a camera facing the operator, allowing the user to take a self-portrait photograph or video while staring at the device's display and frequently displaying a live preview of the image. Independent cameras are directed forward, away from the operator.

TABLE 8.1 Analysis of Hardware and Software of Smartphone and Feature Phone

	Smartphone	Feature Phone
Display	High resolution pictures, rich color range of approx. 16.7 million (~24-bit).	Moderate color range of approx. 4 k–260 k (12-bit to 18-bit).
Memory	Extensive storage capability (~128 GB).	Limited storage capability (~5 MB).
Camera	High-definition image and video capturing. With more advancement, 108 mega-pixel cameras are present in many new models.	The feature phones are generally equipped with the basic camera of approx. 0.3 mega-pixel.
Processor	High speed processor of (~1.5–2.5 GHz octa-core).	Can achieve limited processing speed of approx. 52 MHz–1 GHz.
Input	Touch screen, handwriting recognition, QWERTY-style keyboard.	Numeric keypad, QWERTY-style keyboard.
	AI enabled voice recognition for automatic call dialing.	Automatic voice assistance not provided.
Battery	Li-ion battery, rechargeable, maximum capacity of approx. 5,000 mAh.	Li-ion polymer battery, rechargeable, maximum capacity of approx. 2,000 mAh.
Connectivity	Voice and high-speed data (5G LTE).	Voice and limited data.
Positioning	Enabled with global positioning systems.	Some of the new feature phones are enabled with GPS receiver.
OS	Android, BlackBerry OS, iOS, Symbian, WebOS, Ubuntu, Chrom OS and Windows Phone.	Closed
Voice Over	Voice, Video, VoLte	Voice
Web and Chatting	Direct HTTP, enhanced instant messaging with text and multi-media content.	Via WAP gateway, instant messaging with TEXT, MMS.
PIM (Personal Information Management)	Enhanced phonebook, calendar, and reminder list.	Phonebook, calendar, and reminder list.
Applications	Extensive range of in-built applications.	Minimal supporting applications.

8.2.4 PAGERS

A pager is a wireless communication device that displays or receives alpha-numeric or voice messages (sometimes known as a beeper or bleeper).

Unlike response pagers and two-way pagers, which can only receive messages, one-way pagers can only read them. In the 1950s and 1960s, the first pagers appeared [10], and by the 1980s, they were commonplace. Pagers must be part of a wider network consisting of a number of permanent transmitters (or base stations, in the case of response pagers and two-way pagers) and users with portable pagers in order to function. These networks can be as modest as a single low-power transmitter in a coffee shop or as huge as India with thousands of high-power base stations.

8.2.5 PERSONAL NAVIGATION DEVICE (PND)

PND is a portable electronic device that combines positioning capabilities (similar to a GPS) with navigation capabilities. Certain PNA devices are essentially personal digital assistants with restricted capability that can be unlocked. According to research firm Berg Insight [11], more than 150 million turn-by-turn navigation systems were in use worldwide as of mid-2009, with estimates ranging from 90 million PNDs to 28 million navigation-enabled mobile handsets with GPS. Approximately 35 million factory-installed and aftermarket in-dash navigation systems are included in this statistic.

8.2.6 WEARABLE COMPUTERS

The term "wearable computer" can be defined narrowly or broadly, including smartphones and even regular wristwatches. They could include unique sensors like accelerometers, heart rate monitors, or, for more sophisticated models, electrocardiogram (ECG) and blood oxygen saturation (SpO_2) monitors [12]. We also include innovative user interfaces like Google Glass, a head-mounted optical display that responds to gestures, in our definition of wearable computers. As with the fusion of PDAs and mobile phones into smartphones, it's possible that specialized wearables will develop into generic all-in-one devices. Some wearables have been placed in unconventional locations, but the vast majority are worn on the wrist (e.g., fitness trackers), around the neck (as a necklace), attached to the arm or leg (smartphones when exercising), or as glasses or a helmet (very prevalent) (e.g., on a finger or in a shoe).

8.3 MOBILE OPERATING SYSTEMS

A mobile operating system, similar to a computer's operating system, serves as the foundation upon which other apps can be installed and operate on mobile devices. The mobile operating system (OS) that powers the gadget you buy at the store has already been selected by the manufacturer. The capabilities of your mobile device, such as the thumbwheel, keyboards, WAP, application synchronization, e-mail, text messaging, and more, are determined by the mobile operating system [13]. Which third-party apps you can use on your mobile device is also dependent on its operating system.

Some of the most widely used and well-known mobile operating systems are as follows:

1. **Apple iOS:** iPhone, iPad, iPod touch, and Apple TV are all powered by Apple's iOS mobile operating system [14]. It was not until 2010 that Apple's operating system (formerly known as the iPhone OS) was rebranded to iOS to reflect the expanding range of devices it now supports. In 2015, alongside the release of OS X El Capitan, Apple updated iOS to version 9.
2. **Google Android:** A Linux-based mobile operating system, has overtaken Apple iOS as the dominant mobile platform. Although Android's original source code was made accessible under open-source licenses, Google now develops the mobile operating system in secret until a significant upgrade is ready to be distributed to original equipment manufacturers (OEMs) and the general public [15]. Smartphone and tablet devices powered by Android are produced by a wide variety of companies, including but not limited to Samsung, Sony, Asus, Amazon, HTC, LG, and Google itself.
3. **Windows Phone:** Microsoft's mobile OS is supported by a wide range of devices and is compatible with a wide range of cellular carriers; it was formerly known as the Windows Mobile platform and is now known as Windows Phone [16]. Windows Phone is available on a wide variety of handsets, including those made by Microsoft, Nokia, Dell, HP, Motorola, Palm, and i-mate. To complement its desktop counterpart, Windows 10, Microsoft announced Windows 10 Mobile in late 2015.

Although they are less common now, some of the older generations of mobile operating systems have included: Palm OS, Symbian OS, Linux, and MXI.

8.4 MOBILE TECHNOLOGY

8.4.1 DIVERSE MOBILE TECHNOLOGY FORMS

1. **SMS:** It is an abbreviation for "Short Message Service." It has existed longer than any other SMS service and is the most popular at present. SMS is becoming increasingly popular throughout the world due to its low cost and ease of usage. SMS now refers to any text message delivered to a mobile device and stands for "short messaging service." Typically, messages are limited to 140 characters in length. SMS was originally developed for use with GSM phones but is now compatible with practically every major mobile service provider [17].
2. **MMS (Multimedia Messaging Service):** This which is the usual method for transferring multimedia content, can also be used to convey messages. MMS, unlike SMS, permits the communication of up to 40 seconds of video, in addition to a single image, a slideshow of many images, or voice. The majority of cutting-edge devices will be able to receive and display MMS messages [18]. Typically, the capability to send and receive MMS messages is integrated into the interface for sending and receiving text messages and activates immediately.
3. **3G:** It refers to "third generation," which describes the technology that allows mobile phones to access the internet [19]. The bandwidths and speeds at which information may be sent grow with the arrival of each new technological advancement. In the 1980s, the first of this new generation appeared. The first generation of cell phones was so bulky and heavy that they had to be installed on the roofs of vehicles. The introduction of the second-generation network in the 1990s made it possible to send and receive text messages. This revolutionary breakthrough made possible a safer network and paved the way for the widespread use of 3G and 4G networks today. In 2001, with the advent of 3G connection-based

networks, mobile phone users could finally access the Internet from the palm of their hand. Soon after, cell phones appeared on the scene, compressing a computer's worth of features into something you can hold in one hand. Using a system of phone towers, the signals may travel large distances at a respectable clip of time.

4. **4G:** In the mobile networking industry, 4G refers to the fourth generation of technology, following on from the 3G and 2G networks. Since it is the most advanced technology, most mobile network service providers have adopted it [20]. Despite the restricted capabilities of 3G networks, consumers were able to stream high-definition movies and access the web on their smartphones thanks to 4G network connectivity. Smartphones and tablets are now capable of performing much of the same functions as laptops and desktops. Fourth-generation wireless networks make it possible to maintain high data transfer rates in virtually any location. In 2012, the United Kingdom began offering 4G service. Compared to the older 3G network, premium 4G delivers download rates of roughly 14 Mbps. Downloading a gigabit of data now takes minutes, if not seconds, on a 4G network rather than the hours it took on a 3G network. The speed at which data may be uploaded also increases dramatically with 4G, from under 0.5 Mbps with 3G to over 8 Mbps with 4G and up to 50 Mbps in theory.
5. **Global System for Mobile Technology:** In this context, "GSM" refers to the Global System for Mobile Communication. The Global System for Mobile Communication (GSM) is an open and digital mobile communication cellular technology [21]. It can operate at 850 MHz, 900 MHz, 1,800 MHz, and 1,900 MHz frequencies. It combines FDMA and TDMA characteristics.
6. **Code Division Multiple Access:** Code division multiple access is abbreviated as "CDMA." It is an instance of multiple access and a channel-accessing system. Multiple access refers to the transmission of data simultaneously from multiple transmitters across a single communication channel [22].
7. **Wi-Fi (Wireless Fidelity):** In order to communicate with other computers or mobile devices across a network, we can take

advantage of Wi-Fi, which stands for "wireless fidelity." When using Wi-Fi, data is transmitted through a radio frequency band in a concentric circle. Wi-Fi (Wireless Fidelity) is an umbrella term for wireless networking technologies that aim to provide a Local Area Network (LAN) experience without the use of physical wires [23].

8.4.2 THE USES OF MOBILE TECHNOLOGY

- Mobile devices are becoming increasingly important in the business world, which has helped facilitate remote teamwork. With the advent of mobile technology, groups of individuals were suddenly able to communicate and share files from anywhere.
- Mobile technology is changing the nature of work. Workers are not required to be in the office at all times and can instead do their duties remotely.
- Incorporating mobile technology into your business operations can help you save both time and money. Remote workers regularly save thousands of dollars. With a mobile phone, you won't need to invest in pricey infrastructure like a landline provider. It's more cost-effective to use cloud services rather than more conventional ones. Your business may be more adaptable and productive with the help of technology.
- It is possible that mobile technologies may greatly increase efficiency. It has been estimated that each employee can save up to 7.5 hours per week by integrating mobile applications. With the use of smartphones and other mobile devices, employees may increase their output.
- Using cloud-based services has become increasingly common in recent years. When it comes to storage capacity, mobile applications hosted in the cloud have shown to be superior to their smartphone counterparts.

8.4.3 ADVANTAGES OF MOBILE TECHNOLOGY

- Our social lives have been revolutionized by the availability of numerous apps that allow us to maintain contact with loved ones at any hour. Our smartphones have made it possible to have a

video conversation with anyone, wherever. Furthermore, the portable device provides us with news from throughout the world.
- The convenience of mobile phones in modern life has facilitated many facets of our regular routines. The modern smartphone makes it possible to check traffic conditions in real-time and adjust travel plans accordingly. It also depends on the weather.
- Everything related to gaming may now be found in one convenient location, thanks to the rise of mobile technology. In order to break up the monotony of our workdays or our breaks, we can indulge in our favorite pastimes—listening to music, watching movies or TV series, or even just watching music videos.
- People are scheduling meetings, sending, and receiving paperwork, making introductions, issuing warnings, and applying for jobs all via their mobile phones, among many other lawful uses. Mobile phones are becoming a must for any professional.
- Nowadays, people also use their phones to make payments instead of a traditional wallet. At this time, utilities might be a convenient way to transfer funds to loved ones who live far away.

8.4.4 DISADVANTAGES OF MOBILE TECHNOLOGY

- Mobile phones have become an integral part of daily life for today's families. However, when we don't have to go anywhere, we waste time online becoming total addicts.
- Because of smartphones, people seldom interact face-to-face anymore, instead preferring to share information and ideas via tweets and status updates on social networking sites.
- There is a serious threat of being exposed due to the growing use of mobile devices. Now, information about your home, family, job, and where you live may be easily gleaned from your online social networking profile if one so chooses to read it.
- The value of mobile devices has led to a price increase. Nowadays, individuals spend a considerable amount on cell phones, when that sum of money may be put to greater use elsewhere, such as in the pursuit of higher education or acquisition of other life-improving goods and services.

8.5 GOOGLE'S MOBILE ALGORITHM

This is Google's most recent modification to its mobile algorithm, which was made public on April 21st, 2015, aims to give mobile-friendly websites a boost in smartphone search results. "Mobilegeddon" is the moniker given to the latest update by media outlets and search professionals [24]. In contrast to past Google algorithm modifications like Panda, Penguin, and Hummingbird, the search engine behemoth provided enough notice prior to this significant update. In December 2014, Google first revealed its intentions to alter its mobile algorithm. Google formally revealed the release date in February 2015 and informed news organizations that "they are testing with a new mobile-friendly ranking algorithm." Searches conducted on mobile devices will be the only ones affected by Google's new mobile-friendly algorithm, which will prioritize mobile-friendly pages and demote those that aren't optimized for mobile use. Rankings on mobile devices are the only ones that will be affected by the Mobile-Friendly Algorithm; rankings on desktops and tablets will remain the same.

8.5.1 THE EFFECTS OF MOBILEGEDDON ON MOBILE SEARCHES

It's still unclear how significant an effect the Mobilegeddon update will have. The impact, however, has been predicted to be greater than that of the Panda and Penguin updates. This could suggest that the algorithm change impacts more than 11% of search results. The brand-new algorithm for mobile devices will rank content on a per-page basis. Some of your pages might not work with this upgrade, but others might. The search engine giant, however, views mobile-friendliness as either all or nothing. The Mobile-Friendly Test that Google employs is a real-time process. Google said the rollout of the new algorithm will take a few weeks. Because it could take anywhere from a few days to a few weeks before you see any difference in mobile traffic, it's important to keep a watch on your site's analytics. Although the mobile-friendly mark may take some time to appear in Google's search results, any modifications you make will be reflected instantly.

8.6 CONCLUSION

In conclusion, the concepts and algorithms used in mobile devices play a crucial role in the functionality and performance of these devices. These algorithms enable features such as communication, location-based services, multitasking, power management, security, machine learning, and graphical representation. Understanding these concepts and algorithms is essential for software and hardware developers, researchers, and business leaders who want to create, design, research, or use mobile technology in their field. Mobile devices have become a vital part of our daily lives, and the constant evolution of technology is pushing the boundaries of what these devices can do. Therefore, staying up to date with the latest concepts and algorithms used in mobile devices is crucial to stay ahead of the curve.

KEYWORDS

- **autonomous navigation**
- **heuristic**
- **machine learning**
- **multitasking**
- **power management**
- **robot**
- **sensors**
- **unmanned aerial vehicle**

REFERENCES

1. Srivastava, V. M., & Singh, G., (2013). *MOSFET Technologies for Double-Pole Four-Throw Radio-Frequency Switch* (p. 1). Springer Science & Business Media. ISBN 9783319011653.
2. Rábová, M. (2014). Mobile phone filmmaking as a participatory medium: The case study of 24 Frames 24 Hours.
3. James, R. J., (2001). *Scientific Charge-Coupled Devices* (pp. 3, 4). SPIE Press. ISBN 978-0-8194-3698-6.
4. Williams, J. B., (2017). *The Electronics Revolution: Inventing the Future* (pp. 245–248). Springer. ISBN 978-3-319-49088-5.
5. Matsumoto, K., et al., (1985). A new MOS phototransistor operating in a non-destructive readout mode. *Japanese Journal of Applied Physics, 24*(5A), L323. Bibcode: 1985JaJAP.24L.323M. doi: 10.1143/JJAP.24.L323. S2CID 108450116.
6. Fossum, E. R., (1993). Active pixel sensors: Are CCDs dinosaurs? In: Blouke, M. M., (ed.), *SPIE Proceedings: Charge-Coupled Devices and Solid-State Optical Sensors III* (Vol. 1900, pp. 2–14). International Society for Optics and Photonics.

Bibcode:1993SPIE.1900.2F. CiteSeerX 10.1.1.408.6558. doi: 10.1117/12.148585. S2CID 10556755.

7. *"Camera Sales in 2020 Have Plummeted as Much as 54%"* (2020). PetaPixel. https://petapixel.com/2020/11/06/camera-sales-in-2020-have-plummeted-as-much-as-54/#:~:text=CIPA%20publishes%20both%20the%20volume,%2C%20which%20is%20down%2054%25.
8. *"Feature Phone Definition from PC Magazine Encyclopedia."* www.pcmag.com (accessed on 27 April 2024).
9. Musgrove, M., (2006). *Nikon Says It's Leaving Film-Camera Business.*" Washington Post.
10. *Pager Services to End Tuesday in Japan After 50 Years,"* (2019). Japan Today.
11. *Berg Insight's PND Report 2009.*
12. Barfield, W., (2015). *Fundamentals of Wearable Computers and Augmented Reality.* CRC Press.
13. Okediran, O. O., Arulogun, O. T., Ganiyu, R. A., & Oyeleye, C. A., (2014). Mobile operating systems and application development platforms: A survey. *International Journal of Advanced Networking and Applications, 6*(1), 2195.
14. Novac, O. C., Novac, M., Gordan, C., Berczes, T., & Bujdosó, G., (2017). Comparative study of Google Android, Apple iOS and Microsoft Windows phone mobile operating systems. In: *2017 14th International Conference on Engineering of Modern Electric Systems (EMES)* (pp. 154–159). IEEE.
15. Gandhewar, N., & Sheikh, R., (2010). Google android: An emerging software platform for mobile devices. *International Journal on Computer Science and Engineering, 1*(1), 12–17.
16. Petzold, C., (2010). *Windows Phone 7.* Washington: Microsoft Press.
17. Brown, J., Shipman, B., & Vetter, R., (2007). SMS: The short message service. *Computer, 40*(12), 106–110.
18. Hsu, C. L., Lu, H. P., & Hsu, H. H., (2007). Adoption of the mobile internet: An empirical study of multimedia message service (MMS). *Omega, 35*(6), 715–726.
19. Korhonen, J., (2003). *Introduction to 3G Mobile Communications.* Artech House.
20. Akyildiz, I. F., Gutierrez-Estevez, D. M., & Reyes, E. C., (2010). The evolution to 4G cellular systems: LTE-advanced. *Physical Communication, 3*(4), 217–244.
21. Scourias, J. (1995). Overview of the global system for mobile communications. *University of Waterloo, 4.*
22. Wong, D., & Lim, T. J., (1997). Soft handoffs in CDMA mobile systems. *IEEE Personal Communications, 4*(6), 6–17.
23. Henry, P. S., & Luo, H., (2002). WiFi: What's next? *IEEE Communications Magazine, 40*(12), 66–72.
24. Sirsat, A. D., (2021). Mobilegeddon: Why google forcing websites to go mobile-friendly. *International Journal of Recent Advances in Multidisciplinary Topics, 2*(7), 69–72.

CHAPTER 9

Internet of Things (IoT): The Smart Technology of Autonomous Navigation Systems

ARYAN,[1] KATYAYANI BHARDWAJ,[2] KANISH CHUGH,[3] and R. K. YADAVD[2]

[1]*Department of ECE., J.C. Bose University of Science and Technology, YMCA, Faridabad, Haryana, India*

[2]*Department of ECE, Indraprastha Engineering College, Ghaziabad, Uttar Pradesh, India*

[3]*Department of ECE, Raj Kumar Goel Institute of Technology, Ghaziabad, Uttar Pradesh, India*

ABSTRACT

Before the introduction of the "internet of things" (IoT) in 1999, finding the distance to the moon from Earth wasn't as simple as saying, "Alexa! What is the distance from the Earth to the moon?" or finding out how many items are stored in your warehouse wasn't just a click away. All these tasks were done manually at that time. With the rise of the Fourth Industrial Revolution, various terms such as IoT (Internet of Things), cloud technology, machine learning, and artificial intelligence (AI) have been introduced. The term "Internet of Things" (IoT) is made up of two words: "Internet" and "Things." The term "Things" in IoT refers to various devices such as sensors, software, etc., having unique identities and capabilities to perform actuating, remote sensing, and live monitoring

Advances in Autonomous Navigation Through Intelligent Technologies.
Niharika Singh, Thipendra P. Singh, Brian Azzopardi (Eds.)

of certain data over the Internet. These devices include glucose monitors, wearable devices, credit card-sized computers like Raspberry Pi, etc. With the rise of IoT, concepts such as smart homes, weather monitoring, smart cities, IIoT (Industrial Internet of Things), etc., have come into the picture. The famous author Heather Muzik once quoted, "Life is real, not ideal." Similarly, IoT has faced its fair share of challenges, such as insufficient storage and the use of large amounts of data beyond the scope of traditional computing algorithms. One of the biggest solutions for these problems is cloud-based technology provided by IBM, Microsoft, and various other platforms. It is without a doubt that IoT has made its mark in the world with a US $14.4 trillion industry, with applications in almost all domains, whether it be smart home, smart city, weather monitoring, etc. Integration of Bluetooth technology with Wi-Fi using specialized gateways can process data at lightning speed. Smart cities will not only be crucial in improving productivity and increasing operational efficiencies, but also in reducing management costs to run the city effectively and efficiently. IoT is the need of the hour and setting the future prospects and application areas, it should be implemented on a commercial scale for the ease of human life. In conclusion, IoT has come a long way in the past few years and in this study, we will cover various aspects of IoT that have achieved great popularity in the last decade, leading to comfort, peace of mind, and quality of life.

9.1 INTRODUCTION

The literal meaning of the word "Internet of Things" is the network of physical objects, so it is known as "things" — that is, something which is embedded with sensors, various software, and other technologies that are used for linking with other devices and swapping data with them over the web [1]. Due to embedded features, wireless data transfer, smart sensors, more data means better decisions, the ability to monitor things and track them, using automation to lighten the workload, thus increasing efficiency by saving money and resources and in return ensuring a better quality. According to the survey, there are so many different perspectives and definitions available in the history of IoT. There are a lot of emerging trends in the domain of IoT. Some Internet of Things trends are emerging, making 2021 look like a very exciting year for linking devices and doing more with data. While there was a reasonable fear that COVID-19 would sideline IoT adoption, companies continue to implement projects and unlock their true business potential. From the universality of digital assistants like Echo, Alexa on the home front to the deployment of large-scale devices in manufacturing, digitalization continues.

9.1.1 MORE IOT FOR MORE INDUSTRIES

Industry-specific use cases will accumulate and give rise to more industry-specific IoT identities: Internet of Medical Things, Internet of Retail Things, Internet of Logistics Things, Internet of Workforce Management, etc. [2] as more businesses find value in data and analytics.

9.1.2 IDENTICAL TWINS WILL BE BORN

IoT efficiency will increase as companies discover the value of hiring the replica of the IoT elements. A digital twin strategy enables a business to create a digital representation of an IoT process, product, or service and use real-time data to monitor, test, and improve the products or processes – the original twin. Using twine without interrupting doing its job.

9.1.3 MOVING PAST DATA

The focus is shifting to processing data quickly for quicker decision-making and its post-processing. IoT is bonding with AI and ML to rapidly process vast amounts of data swiftly [3]. Companies are moving previously assembled information (which was in a simple version) to synthesize it to make smart and informed recommendations and decisions. IoT and machine learning were the only technologies that people heard about emerging to make people's lives simpler. But in recent years, these technologies have successfully made it into providing applications that can provide users with a smarter, simpler, reliable, and faster way to perform all tasks and a more straightforward workflow in all areas, increasing productivity. In the recent rise of popularity of these technologies, Internet of Things (IoT) or Machine Learning (ML) has an easy way to do just about every possible task effectively and efficiently [4]. Artificial intelligence (AI) is becoming increasingly polished as computing power advances, the knowledge of data scientists, the resource sharing, and the proper availability of machine learning tools to create advanced algorithms; the day is not too far when the Internet of Things will become a mainstream event or phenomenon.

9.1.4 MORE WILL BE DONE AT THE EDGE

IoT data processing solutions at the edge give us the option of faster decision-making. The growth in 5G networks and connected devices is driving digitalization by bridging the gap between device and cloud and enabling more analysis before it gets there [5]. These are some of the trends in IoT development that we can see, and more are likely to come out and make the near future an exciting time for IoT [6] (Figure 9.1).

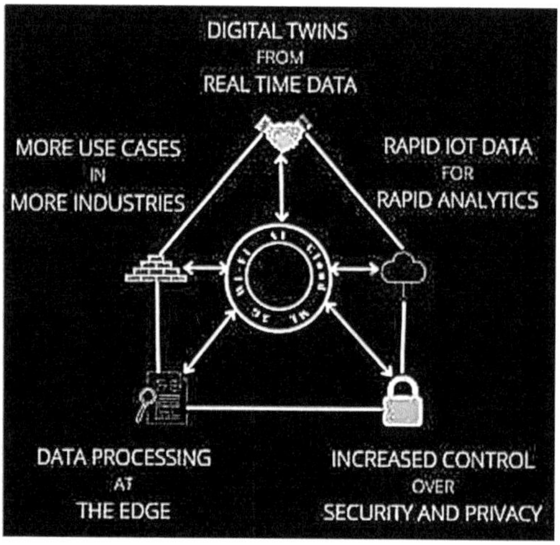

FIGURE 9.1 The trends in IoT.

Source: Original and Own work.

9.2 CONTRIBUTION IN ACHIEVING SUSTAINABLE DEVELOPMENT GOALS (SDGS)

The Sustainable Development Goals (SDGs) are a call to action by all major countries to promote welfare and at the same time protect the planet. In addition to tackling climate change and advancing environmental protection strategies, to end poverty, we must encourage economic growth and meet a range of social needs, including social security, health, education, job opportunities, and so on [7]. Labeling values and encouraging the priority of sustainability goals to enhance the powerful impact IoT can provide for a better and sustainable world. Nowadays, IoT technology is emerging to improve the quality of life of citizens and the competitiveness of companies. Reasons to invest in IoT technology:

1. Productivity is affected by 40% due to the existing constraints in the infrastructure in low-income countries.
2. Hunger is a very big issue in such a developed world. The statistics of it are really very depressing. According to a report by the World Food Program, one out of every nine persons in the world is facing hunger. We can use IoT to resolve this through different methods. Food wastage is also a major issue. If we combine them both, we can overcome hunger and food wastage easily. This can be effective in ways like improving the yield from harvesting and storage, and also by enhancing the existing distribution network. We can create a 24/7 monitoring system for preserving food and generating an alert system whenever action is required by using sensors to indicate higher temperatures. This step can help us take preventive measures at the right time to reduce food wastage. Food wastage can be minimized by contacting nearby stores, hotels, and restaurants to give/donate food to local charities and other distribution programs instead of throwing it away. IoT can be used here by disseminating information in real-time to each stakeholder, thus making sharing more efficient. According to reports, about 821 million people were malnourished in 2017. In order to feed 9.1 billion people, annual grain production would need to increase to about 3 billion tons and annual meat production would need to increase from the existing 200 million tons to 470 million tons by the next three decades, that is 2050 [8].
3. Worldwide, billions of people have no access to basic sanitation. They have communicable diseases which will cost low and middle-income countries more than 7 trillion in the next 15 years.

In 43 countries, about 700 million people are suffering from water shortage. By 2025, over 1.8 billion people will be living in areas or cities with absolute water shortages, and 2.3% of the population could be living under water stress conditions.

Earth's population is 7 billion as of now. Half of the population, about 3.5 billion people, live in cities, and over 4.5 million people are projected to live in cities by the next decade. World cities occupy 3% of the earth's land but contribute to 60–80% of energy consumption and 75% of carbon emissions [9].

According to current emissions and concentrations of greenhouse gases, there is a possibility that the average temperature of the earth will exceed 1.5°C by the end of this century. Since 1990, global carbon dioxide emissions have increased by almost 50% (Figure 9.2).

FIGURE 9.2 Various SDGs and their relation with IoT.

Source: Original and Own work.

9.3 APPLICATIONS OF IOT

9.3.1 *SMART PILLS*

Orally deployable electronic medical devices shown in science fiction movies that we all dreamed about are not a part of the future but a thing in the present. Biomedical, pharmaceutical, and food industries have been researching edible tech for decades now. One of the biggest advancements in the medical industry worldwide over the years [10]. These smart pills ensure better diagnostics, real-time monitoring, and appropriate treatment of various life-threatening medical conditions by granting medical professionals access to monitor the patient's body functions in real time. The net worth of the global smart pills industry is about $779.9 million, which is expected to reach an all-time high of $3 billion by 2025.

9.3.2 SMART WEARABLES

In the past few decades, the closest thing to a smart wearable was a Casio calculator watch. No one knew that these wearables could emerge as an element of intense business and social change as well as be a lifesaving development [11]. The wearable devices are the omnipresent technology of the Internet of Things (IoT) having various applications in day-to-day life. The devices such as smart wrist wear, smart clothes, and medical wearables are currently concentrated on the health, safety, interaction, tracker, identity, fitness aspects which makes the popularity grow even more than having a piece of what used to be science fiction in the 90s become reality on your wrist. These wearables can not only be used to monitor but also can be used to control your other smart devices.

9.3.3 ROBOTICS

With recent developments in nanorobotics, the deployment of miniature-sized internet-controlled and connected robots inside a human body has become quite trouble-free and painless. These small robots can perform surgery that would be inconvenient otherwise and reduce the incisions required to perform any surgery, thus leading to a less invasive process and faster curing and healing for patients [11]. Moreover, IoT is making huge leaps forward in sectors like prosthetics, elderly care, managing disease outbreaks (COVID-19), soft robotics, and rehabilitation. It is quite clear that IoT, when mixed with robotics, is going to shape the future.

9.3.4 SMART LIGHTING

Having the power to control lights using your voice or remotely from the comfort of your bed was something only a god could have, but in these times that is something anyone could do and feel all-powerful [13]. Smart lighting is just like your regular lighting but with the option to control it from an app on your phone using Bluetooth, Wi-Fi, or over the internet. These devices enable you to change the brightness or color of the light depending on what kind of smart light you have [14]. This opens up a lot of opportunities for getting full control depending on your mood and different settings for different times.

9.3.5 SMART SECURITY

In these developing times, security concerns are also on the rise, making it necessary in today's world. A smart security system ensures that your property is protected by a number of intelligent devices [15]. This 2.49-billion-dollar industry has changed the way we think about home security. There is a statistic that gives us an overview that 88% of all theft-related crimes are residential in nature. Furthermore, there is a 300% possibility of theft in homes without these smart security systems.

9.3.6 SMART PLUG

We all know the feeling when we think that we forgot to switch off an appliance or the feeling when you have to get up to switch off a light or to find out energy consumption [16]. In today's world, electricity is consumed in almost everything that we do. The challenge was to come up with new and innovative ideas and solutions that would monitor and manage electric power usage and provide wireless control over the plugs using relays (electronic switches). This will turn your appliances into smart devices by connecting them to a smart socket, allowing you to control them from anywhere, schedule the appliances to turn them on/off, and monitor the energy consumption and status of the appliances.

9.3.7 SMART SPEAKER

Nearly 90 million U.S. adults' own smart speakers, which is 32% more than a year ago. So, what is a smart speaker? It is a type of speaker that has an integrated virtual assistant that provides interactive actions and can be synced with all the IoT devices mentioned above, controlling all those devices from a single device [17]. A smart speaker is a device that helps you with everyday tasks like creating a shopping list, getting a recipe, or even something like turning off your air conditioner with just a voice command.

9.3.8 SMART METER

Have you ever thought that your meter readings were inaccurate or wanted to check the energy usage of your entire home [18]? Well, a few years ago, you might not have even considered if it was possible or not. However, in recent years, with the development of smart metering, a smart meter is a device that monitors electricity consumption, provides you with accurate bills, tracks your energy usage, and delivers them to you remotely over the network, saving you a lot of money if used correctly [19].

9.3.9 SMART GRID

Ever wondered what the next generation of energy systems would be? The "GRID" is an electricity network enabling a two-way flow of electricity and data, making use of digital communication technology which enables the providers to firstly detect, then react, and finally pro-act upon the changes and issues, having full control over each and every connection [20]. A smart grid has automatic control over every aspect of the connection in order to deliver sustainable, economic, and secure supply of energy. The main association of a smart grid is with the production of energy using renewable resources [21, 22].

9.3.10 SMART TRAFFIC AND STREET LIGHTS

With the development of internet-connected cars capable of connecting to the internet using inbuilt systems, it is a lot easier to connect to the network. But the question arises: how does it help the traffic [23]? With the fast growth of population, traffic monitoring has become a great challenge. Internet-connected cars manage it in real-time while being connected to both central and local servers, providing data using ultrasonic, GPS, and also data gathered from traffic feeds, mobile devices, road cameras, weather stations, etc., which could also be gathered by smart street lights containing technology such as cameras, light-sensing photocells, and other sensors connected to a server over wired or wireless networks [24].

9.3.11 SELF-DRIVING CARS

It is a vehicle that makes use of a combination of cameras, sensors, radar, actuators, and deep learning algorithms to utilize the data gathered by the car and give instructions to the car, which will navigate without any human intervention to a destination that has already been determined over the roads [25]. Companies like Tesla, Google, Volkswagen, Audi, and BMW have launched their self-driving cars into the wild. The neural networks act as an identifier by identifying patterns in the data, which run on machine learning algorithms. The data includes images captured from the cameras embedded on self-driving cars, which can significantly reduce the number of crashes and help with economic activity, saving lives and money [26] (Table 9.1).

TABLE 9.1 Summary of Applications of IoT

IoT Application	Description
Smart pill	Small ingestible electronic devices used for monitoring a person's health.
Smart wearables	Wearable technology is an electronic device that can be worn as accessories.
Robotics	Devices that can monitor events and mix sensor data from different sources.
Smart lighting	Lights that allow users to manage its properties in real time using their smartphone.
Smart security	Number of intelligent devices that together make a place secure.
Smart plug	Electrical outputs that let you automate your electronic devices.
Smart speaker	A speaker that helps you with everyday tasks with voice assistants.
Smart meter	A device which monitors electricity consumption and provides you with accurate bills and tracks your energy usage.
Smart grid	Electricity network enabling a two-way flow of electricity and data.
Smart city	Cities using IoT devices for better functioning.
Smart traffic light and streetlights	Internet connected traffic lights and streetlights having multiple sensors for data collection.
Self-driving cars	A vehicle that uses multiple sensors for navigation without human interference.

9.3.12 REAL CASE STUDY OF BARCELONA AND SINGAPORE

In the past few years, Singapore and Barcelona have become beacons of smart countries and cities with the integration of IoT at their core by launching their smart nation vision. Ideas like digitized healthcare systems, which include smart elderly alert systems and chatbots that talk to the elderly, have made a huge impact on the elderly, and the implementation of self-driving shuttles has also improved the mobility of the people. It is without a doubt that these new inventions and implementations are going to pave the way for the future [27].

9.4 FUTURE SCOPE AND CONCLUSION

Recent advancements in the domain of IoT have made it a center of attraction for researchers and developers worldwide. In this study, we have reviewed the concept of IoT, establishing a correlation between IoT and SDGs, and doing a detailed analysis of various IoT applications [28]. With the extensive growth and expansion of interconnected sensors and devices, the technology of IoT is increasing rapidly. It aims at establishing smart industries by transforming the existing ones into intelligent ones by utilizing the dynamic network of interconnected devices and providing a smart output. A huge amount of streaming data is generated by these IoT devices which needs effective storing and needs further analyzing for decision making in real time. Therefore, its continuous expansion giving rise to complex issues such as management, collection, storing, processing, and analyzing the data. IoT researcher and developers are working hand in hand to extend this technology on larger and commercial scale by overcoming these challenges and benefiting the society in the best possible manner. Various application areas of IoT have also discussed where IoT developers, engineers and researchers are engaged [29]. In a collaborative manner, IoT and Industry 4.0 can provide an extended scope of some contemporary solutions, applications, and services that can yield significantly on personal, professional, and economic level by providing enormous opportunities [12]. Although, the offerings of IoT on the grounds of solutions to organizations and industries as well as advanced applications and services are enormous, but at present, it is still at an early stage of development, deployment adoption and implementation in the society;

thus, facing various open challenges and issues that still need to be thought upon and addressed.

KEYWORDS

- **decision making**
- **dynamic network**
- **internet of things**
- **IoT applications**
- **smart technology**
- **sustainable development goals**
- **trends in IoT**

REFERENCES

1. Balaji, S., Nathani, K., & Santhakumar, R., (2019). IoT technology, applications and challenges: A contemporary survey. *Wireless Personal Communications, 108*(1), 363–388.
2. Jennath, H. S., Adarsh, S., & Anoop, V. S., (2018). Distributed IoT and applications: A survey. *Integrated Intelligent Computing, Communication and Security*, 333–341.
3. Nath, M. P., Priyadarshini, S. B., Mishra, D., & Borah, S., (2020). A comprehensive study of contemporary IoT technologies and varied machine learning (ml) schemes. *Advances in Intelligent Systems and Computing*, 623–634.
4. Greco, L., Percannella, G., Ritrovato, P., Tortorella, F., & Vento, M., (2020). Trends in IoT based solutions for health care: Moving ai to the edge. *Pattern Recognition Letters, 135*, 346–353.
5. Wang, D., Chen, D., Song, B., Guizani, N., Yu, X., & Du, X., (2018). From IoT to 5G I-IoT: The next generation IoT-based intelligent algorithms and 5G technologies. *IEEE Communications Magazine, 56*(10), 114–120.
6. Chettri, L., & Bera, R., (2020). A comprehensive survey on internet of things (IoT) toward 5G wireless systems. *IEEE Internet of Things Journal, 7*(1), 16–32.
7. Oliva-Maza, L., Torres-Moreno, E., Villarroya-Gaudó, M., & Ayuso-Escuer, N. (2019). Using IoT for sustainable development goals (SDG) in education. *Multidisciplinary Digital Publishing Institute Proceedings, 31*(1), 1.
8. Lopez-Vargas, A., Fuentes, M., & Vivar, M., (2020). Challenges and opportunities of the internet of things for global development to achieve the United Nations sustainable development goals. *IEEE Access, 8*, 37202–37213.

9. Sanchez, D. O., (2019). Sustainable development challenges and risks of industry 4.0: A literature review. *Global IoT Summit (GIoTS).*
10. Pedi, R. J. E., & Chavan, A., (2020). AI-IoT based smart pill expert system. In: *2020 4th International Conference on Trends in Electronics and Informatics (ICOEI),* (48184).
11. Verma, V., Chowdary, V., Gupta, M. K., & Mondal, A. K., (2018). IoT and robotics in healthcare. In: *Medical Big Data and Internet of Medical Things* (pp. 245–269). CRC Press.
12. *Internet of Things Declaration to Achieve the Sustainable Development Goals,* (n.d.). IoT Week. https://Iotweek.org/internet-of-things-declaration-to-achieve-the-sustainable-development-goals/ (accessed on 27 April 2024).
13. Sikder, A. K., Acar, A., Aksu, H., Uluagac, A. S., Akkaya, K., & Conti, M., (2018). IoT-enabled smart lighting systems for smart cities. *IEEE 8th Annual Computing and Communication Workshop and Conference (CCWC)* (pp. 639–645).
14. Adnan, N., Kamal, N., & Chellappan, K., (2019). An IoT based smart lighting system based on human activity. *IEEE 14th Malaysia International Conference on Communication (MICC)* (pp. 65–68).
15. Somani, S., Solunke, P., Oke, S., Medhi, P., & Laturkar, P. P., (2018). IoT based smart security and home automation. In: *4th International Conference on Computing Communication Control and Automation (ICCUBEA)* (pp. 1–4).
16. Oh, J., (2020). IoT-based smart plug for residential energy conservation: An empirical study based on 15 months' monitoring. *Energies, 13*(15), 4035.
17. Mocrii, D., Chen, Y., & Musilek, P., (2018). IoT-based smart homes: A review of system architecture, software, communications, privacy and security. *Internet of Things, 1, 2,* 81–98.
18. Abate, F., Carratù, M., Liguori, C., Ferro, M., & Paciello, V., (2018). Smart meter for the IoT. *IEEE International Instrumentation and Measurement Technology Conference (I2MTC)* (pp. 1–6).
19. Bin, R. A., & Kotb, Y., (2020). Towards a real IoT-based smart meter system. In: Luhach, A., Kosa, J., Poonia, R., Gao, X. Z., & Singh, D., (eds.), *First International Conference on Sustainable Technologies for Computational Intelligence: Advances in Intelligent Systems and Computing* (Vol. 1045). Springer, Singapore.
20. Al-Turjman, F., & Abujubbeh, M., (2019). IoT-enabled smart grid via SM: An overview. *Future Generation Computer Systems, 96,* 579–590.
21. Barman, B. K., Yadav, S. N., Kumar, S., & Gope, S., (2018). IoT based smart energy meter for efficient energy utilization in smart grid. In: 2nd *International Conference on Power, Energy and Environment: Towards Smart Technology (ICEPE)* (pp. 1–5).
22. Mugunthan, S., & Vijayakumar, T. (2019). Review on IoT based smart grid architecture implementations. *Journal of Electrical Engineering and Automation, 10*(1), 12–20.
23. Dheena, P. P. F., Raj, G. S., Dutt, G., & Jinny, S. V., (2017). IoT based smart street light management system. *IEEE International Conference on Circuits and Systems (ICCS)* (pp. 368–371).
24. Das, A., Dash, P., & Mishra, B. K., (2018). An innovation model for smart traffic management system using internet of things (IoT). In: Sangaiah, A., Thangavelu, A., & Meenakshi, S. V., (eds.), *Cognitive Computing for Big Data Systems Over IoT;*

Lecture Notes on Data Engineering and Communications Technologies (Vol. 14). Springer, Cham.

25. Geetika, M., et al., (2019). A study on self-driving car an application of IoT. *International Journal of Computer Networking, Wireless and Mobile Communications, 9*(1), 25–34.
26. Marosi, A. C., Lovas, R., Kisari, A., & Simonyi, E., (2018). A novel IoT platform for the era of connected cars. *IEEE International Conference on Future IoT Technologies (Future IoT)* (pp. 1–11).
27. Rayes, A., & Salam, S., (2018). Internet of things (IoT) overview. *Internet of Things from Hype to Reality*, 1–35.
28. Maheswar, R., & Kanagachidambaresan, G. R., (2020). Sustainable development through internet of things. *Wireless Networks, 26*(4), 2305–2306.
29. Chaudhuri, A., (2019). *Internet of Things, for Things, and by Things*. CRC Press.

CHAPTER 10

Issues in Autonomous Navigation

DEEPTI SINGH, ANKITA VERMA, and ARPITA JADHAV BHATT

Department of CSE&IT, Jaypee Institute of Information Technology, Noida, Uttar Pradesh, India

ABSTRACT

Autonomous navigation provides the ability to a vehicle to determine its location and plan a path to reach some pre-decided goals without any human intervention. Here, a vehicle could be a car traveling on some road, an unmanned aerial vehicle in airspace, a spacecraft traveling in the universe, submersible collecting information from the depth of the ocean. There is a plethora of applications in different sectors like transport, agriculture, healthcare, e-commerce, space industry where the autonomous navigation system can make a substantial contribution. As this is an emerging field with potential applications in varied domains, in-depth analysis of the basic architecture, applications, and its associated challenges becomes crucial. To this end, in this chapter, we will discuss the major issues that are still open in the autonomous navigation system such as the development of efficient sensors and actuators, novel routing and interaction protocols enhancing security of the system, path planning and following, mapping, and localization to name a few.

10.1 INTRODUCTION

In today's world, the advancement in technology has been at a rapid pace, enabling major and faster changes in our day-to-day living. This is due to the increase in the usage of internet-enabled gadgets producing

Advances in Autonomous Navigation Through Intelligent Technologies.
Niharika Singh, Thipendra P. Singh, Brian Azzopardi (Eds.)

enormous data and the competitiveness in businesses to offer their products and services in the market in the best possible way. Some of the leading technologies include Artificial Intelligence & Machine Learning, Big Data Analytics, Blockchain, Robotics, Quantum Computing, etc. Human intelligence is behind all these technological advancements, but we still require Artificial Intelligence (AI) to perform tasks in which humans lag behind, such as making unbiased decisions, working continuously without fatigue, executing multiple tasks in a very short span of time with more accuracy, and many more [1]. AI has given rise to Autonomous Technology (AT), enabling autonomous systems with human-like intelligence and all the benefits of machine-led tasks. After the COVID-19 pandemic, there has been a significant demand for such autonomous systems that require minimal to no human interaction. Some examples include Drone Delivery Systems, Chatbots, Driverless Taxis, Robots for Clinical Diagnosis, etc. Autonomous Navigation Systems (ANS) are one of the by-products of these technologies, which have garnered significant interest from the automobile and robotics industry. Navigating these vehicles with minimal human intervention is the key feature of ANS. One of the popular examples of ANS is a self-driving car by the well-known automotive US company Tesla. These self-driving cars offer various features like fully automated, semi-automatic, or partially automated, depending on the level of autonomy described in Section 10.3 in more detail. ANS relies on other technologies such as sensors, cameras, Global Positioning System (GPS), learning the environment features, estimating the current state and localization, tracking objects, path planning, and then following the planned path to reach the destination [2].

10.1.1 WHAT IS AUTONOMY?

In simpler and generic terms, autonomy means to be controlled or function by one's own self. In terms of technology, autonomy can be understood as a technology which performs the given task without any human control. Autonomy is helpful in many ways for our society; it could help us in doing monotonous tasks which humans might not be interested in doing after a time [3]. Also, sometimes we don't have enough workforce to do these tasks; machines are much faster in accomplishing these tasks as compared to humans. Through human intelligence, we build autonomous systems/devices which have huge applications in areas like medical, transport, logistics, defense, science & space, agriculture, e-commerce, education, etc.

10.1.2 WHAT IS A NAVIGATION SYSTEM?

Navigation is a process that is used to direct a vehicle from one place to another, covering all the sub-processes from vehicle monitoring to controlling its movement. A navigation system is a device capable of identifying the current location and helping reach the destination on a given map. Most popular navigation systems are GPS-based, which use orbital data from satellites to calculate the location of the object on the map and navigate along the path to reach another point [4]. Navigating from one place to another includes some generic steps starting from pre-processing, which includes tasks like identifying the surroundings and current position, planning a proper path on a given map, and then acting upon that path to reach the goal while avoiding all obstacles in the path.

10.1.3 WHAT IS AUTONOMOUS NAVIGATION?

Autonomous navigation systems are those systems built with the capability of defining a route in the given environment and then following it to reach a goal position by utilizing self-intelligence, also avoiding collisions in case they encounter any in the route. There are six levels of automotive automation as per Society of Automotive Engineers (SAE) standards [5], mentioned in Figure 10.1. As the number of levels increases from 0 to 5, the system will include more automation tasks. In Level 5, the system is fully automated, and human interference is not needed in fully autonomous navigation systems. The key feature of autonomous navigation systems is their ability to make decisions by themselves regarding the navigation plan. These systems are embedded in many vehicles such as agricultural robots, flying drones, cars, ships, battle tankers, etc.

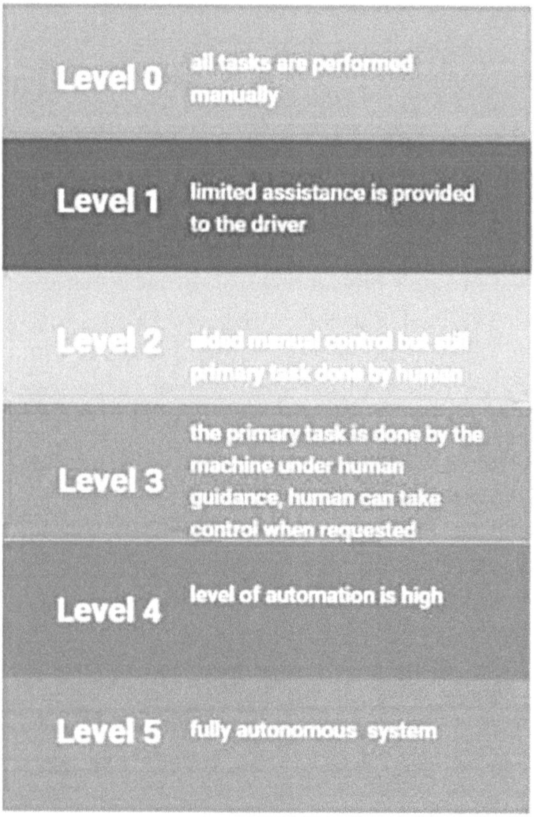

FIGURE 10.1 Levels of automotive automation.

10.1.4 TYPES OF NAVIGATION SYSTEMS

There have been a lot of navigation systems designed in the industry so far based on different navigation techniques proposed by the researchers. There are a lot of categories of navigation systems available in the literature, some are dependent on the type of environment these navigation systems are supposed to be deployed – indoor or outdoor, some are simply categorized based on the technology being used in making these systems like GPS, Radio-frequency identification (RFID), sensor, camera, inertial navigation. Other categorizations are done based on autonomy like whether the system is automatic, partially automatic, semi-automatic, etc. Another category is based on the place where the navigation system has to navigate – land or road, aerospace, marine, and aeronautic. These navigation systems are also designed according to their usage, for example, a navigation system used in agriculture is specifically designed as per the agricultural needs [6], in space the navigation system has to use different sensing techniques to sense the surroundings than in marine, in surgical procedures the navigation system has a different aim and so on.

10.2 RELATED WORK

Miranda et al. [7] proposed an autonomous navigation framework for drones performing delivery jobs. The proposed framework ensures completion of delivery tasks by these drones in densely populated urban areas although they have assumed that the environment is obstacle-free. Also, this navigation framework relies on the availability of GPS which is yet another limitation of this model. In a study, Joo et al. [8] worked on the novel navigation framework using the capabilities of semantic approaches for modeling, information processing, and navigation for robots. They have used TOSM (Triplet Ontological Semantic Model) for understanding the environment features for navigation. In another study, Polvara et al. [9] discussed the obstacle avoidance techniques based on Ref. [9] used in autonomous vehicles. They have compared various sensors used in sensing the surroundings of a vehicle and discussed their accuracy for collision detection techniques. They have also mentioned a set of standard international rules for avoiding collision at sea in varied scenarios. Trajectory generation in a frequently changing environment is also a very challenging issue in ANS. Wang et al. [10] have proposed a novel framework for trajectory generation and path chasing in aerodynamic disturbing and regularly fluctuating atmosphere. In another study by Pedersen et al. [11], the issue of verification of ANS is highlighted. Testing and verifying in a real-life setup are a very costly and time-consuming process. A prototype of a test system including testing using simulators, integrating the ANS and simulator are discussed in this chapter.

10.3 KEY ISSUES IN AUTONOMOUS NAVIGATION

ANS is for sure a challenging field of study, making any device decide the right path to reach another point and pass all the obstacles in the path takes a lot of intelligence. It includes various issues ranging from path planning, sensing the environment to make ANS aware of the current map and localization of the ANS device with respect to the environment it is functioning in, following the trajectory, nature of the environment – indoor, outdoor, static, or dynamic, etc. Also, if the ANS is supposed to function in an outdoor system, then it is heavily influenced by weather conditions, radar functioning, type of land, aerospace, marine, and aeronautics. ANS is definitely going to revolutionize various other industries like agriculture, defense, transportation, healthcare, etc., in the coming years, but we need to work on the challenges faced during various phases of it [12]. In Figure 10.2, we have shown the most common issues with the amount of work done on each of these issues in our literature. Each of these issues comprises many other factors which we are going to discuss one by one in the following sub-section.

10.3.1 PATH PLANNING

For autonomous navigation systems, path planning is one of the crucial steps to navigate from source to destination. The best possible path for such systems is a trade-off between various parameters like total time to reach the destination, number of obstacles faced in the route, algorithm used in path planning, etc. [13, 14].

10.3.2 MOTION CONTROL

Motion control is another crucial issue in designing ANS. Based on the trajectory identified through any suitable trajectory algorithm, the ANS device is supposed to run the acceleration, brake, and steering commands by itself. These commands need to be very precise for a fully automated navigation system; however, in practice, increasing precision makes the ANS very expensive. Additionally, there are temperature, vibration, and contamination-related issues that hinder the motion control strategies used in ANS [15].

10.3.3 STATE ESTIMATION

ANS designed for underwater, aerospace, and aeronautics. It is very crucial to identify the current state measurement of the ANS device, as it will also

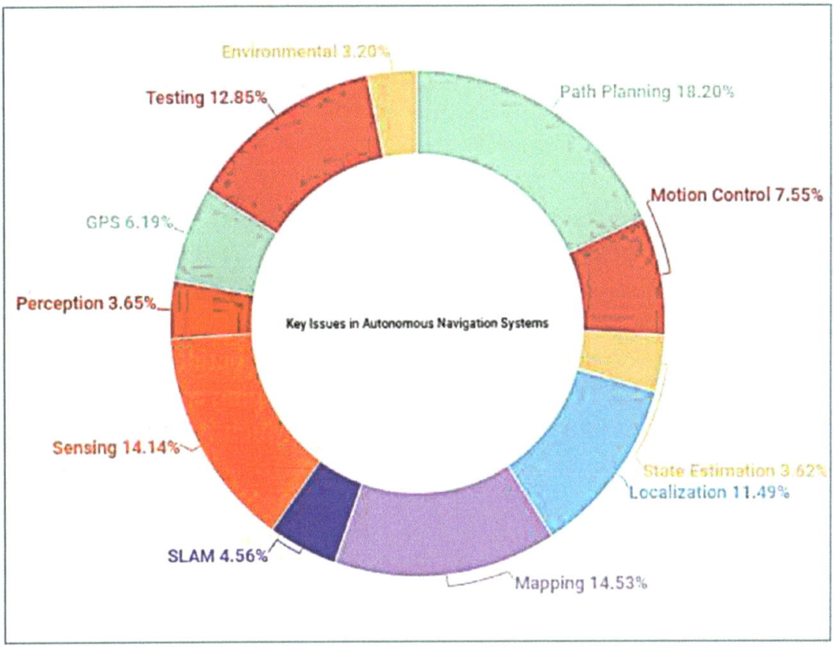

FIGURE 10.2 Key challenge areas in ANS.

affect the accuracy of other steps – localization and trajectory planning of ANS framework. In a dynamic environment, it is a very complex task to estimate the current state of the ANS device. Various techniques like Kalman filter and Information filter are used in the literature for state estimation problem [16].

10.3.4 LOCALIZATION

Localization in ANS is used to identify the current position of autonomous navigation vehicle/vessel with respect to a known point in the environment map. Precise localization is very much needed for the efficient functioning of ANS in the current map. The major challenges to compute precise localization are the complexity involved in computation tasks, delays in communication, and noise in sensor readings. This will also affect other steps of ANS framework like path planning and perception of the present environment [17]. Also, in outdoor environments, due to harsh climatic conditions, it becomes very difficult to localize the navigation object.

10.3.5 MAPPING

Mapping algorithms are used to map the ANS device to its environment in which the device is currently functioning in Ref. [18]. In mapping algorithms for autonomous navigation devices, there is a huge dependency on good quality sensors which help in making the grid map of the environment. Another major issue is to update the map continuously as per the changing dynamics of the environment. The mapping consists of keeping track of each coordinate in the environment, which is yet again a time-consuming and complex task.

10.3.6 SLAM (SIMULTANEOUS LOCALIZATION AND MAPPING)

The state space created by appending the autonomous navigation device position and landmark locations in SLAM is fully coupled, which makes full observability difficult. Another issue with the fully correlated Kalman filter-based method to SLAM is its computational complexity, which is due to the Kalman approach's computation of covariance matrices whose size increases quadratically with respect to the size of the map. This also leads to a large response time [19].

10.3.7 SENSING

To sense the current environment, various sensors are used in ANS systems. These sensors are one of the primary components of ANS devices/vessels. The most popular sensors used are LiDAR (light detection and ranging) sensors, radar sensors, SONAR sensors, cameras for imaging, IR sensors, GPS sensors, gyroscopes, and magnetometers. These sensors face difficulties due to bad light and weather, interference from magnetic fields, and high reflection. Also, some are extremely expensive to be deployed in ANS devices. Sensor integration with a complete end-to-end ANS system is also one of the complex issues [20].

10.3.8 PERCEPTION

The perception for ANS involves the process of integrating four tasks: the destination points to be reached, the locations of static obstacles, the present and future locations of moving obstacles, and the vehicle's current speed, position, etc. Data about the environment is acquired using various sensors, and then data fusion is performed to centralize the data to be fed into the ANS framework design. Varied fields of view need to be considered while capturing the data from these sensors for an efficient perception mechanism [21].

10.3.9 GLOBAL POSITIONING SYSTEM (GPS) USAGE

GPS dependency is also one of the challenges for ANS-enabled vehicles which solely depend on the accuracy of GPS signals to navigate in the current map. GPS signals are not available at all places, for example, in some rural areas, low-density mountain areas, Arctic regions, etc. Many times, the GPS signals lag behind the time when they were actually required by the ANS device, leading to large response time and communication delays [22].

10.3.10 TESTING

Testing and validation of ANS-enabled devices/vehicles is one of the key challenging areas. It is quite difficult to recreate the critical real-life scenarios in practice. Also, simulators have limited functionalities to simulate all possible scenarios of real life. Simulators must be feature-rich to provide full control over the virtual environment considering scenarios of traffic, terrain, weather, low-density areas, etc. [23]. Sufficient test beds should be deployed to test and validate these ANS vehicles. Also, all boundary cases must be tested for these systems. Testing at the vehicle level will not be sufficient to ensure safety. It has long been recognized that testing systems thoroughly enough to ensure ultra-reliable system functioning is impossible.

10.3.11 ENVIRONMENTAL

An indoor environment is a completely different setup than the outdoor environment for an ANS to function upon. The indoor environment is less prone to harsh weather conditions compared to the outdoor environment. Likewise, we have a static environment in contrast to a dynamic environment which is more challenging for an ANS to provide an efficient outcome [29]. For example, if we take an example of a self-driving car running on a road, there are other moving vehicles on the same road which need to be taken care of by the ANS. An example in a static scenario is a simple home-cleaning robot system working in a closed room with all non-moving things in the room. ANS designed for aerospace will be different from the one designed for a marine or aeronautics environment. ANS designed to be used in agriculture fields have to consider the unevenness in the piece of land.

10.4 INDUSTRY SPECIFIC CHALLENGES

This section will discuss some of the challenges that arise due to the application of ANS in varied sectors such as transportation, urban travel, agriculture, logistics, defense, healthcare, and so on.

- One of the most pressing problems for Self-Driving Vehicles (SDVs) is their safety. In SDVs, determining the acceptable degree of risk becomes critical. The focus should be on mitigating some of the novel hazards such as unintentional system failure, cyber-attacks, data breaches, and so on. Thus, their commercialization will necessitate regulatory approvals and extensive testing [24].
- In order for autonomous navigation systems to thrive in the transportation domain, they need a lot of help from infrastructure. Focusing on the development of cost-effective technical solutions that encourage optimal resource utilization is crucial [25].
- For the applicability of ANS in varied domains, the effectiveness of the sensors employed in acquiring relevant perception from the environment and the decision of which sensor or combination of sensors would be best in the specific case is critical [26].
- In the agricultural domain, the guided navigation system has to deal with a variety of vegetations, terrains, landscapes, and weather conditions [27].
- Dedicated infrastructure, such as complicated robot mechanisms, networking resources, human-computer interfaces, and a technologically sound workforce, are necessary to actually realize the

full potential of automation. For developing and underdeveloped countries, facilitating these is a monumental task.
- The public's satisfaction with autonomous systems in the healthcare sphere appears to be mixed and ambiguous. Policymakers must endeavor to create systems that are extremely precise [28].
- In e-commerce, autonomous systems must be built with the ability to sustain and repair human collaborator relationships.
- In supply chain management, ANSs are usually applied for repetitive simplistic tasks, their application in complex decision-making tasks with incomplete and uncertain knowledge is still limited [25].

10.5 CONCLUSION

In this chapter, we have highlighted some critical issues and research challenges in the domain of autonomous navigation systems. From the state of the art, it is evident that academia as well as industry are truly dedicated to the research and development activities in this domain with the aim to achieve full autonomy in performing mundane, risky, time-consuming, and challenging tasks. However, in order to expand the outreach of the ASNs in the global market and integrate their applicability in day-to-day life, proper infrastructure and cost-effective solutions are the need of the hour.

KEYWORDS

- **autonomous navigation**
- **cost-effective solutions**
- **e-commerce**
- **heuristic**
- **policymakers**
- **robot**
- **sensors**
- **unmanned aerial vehicle**

REFERENCES

1. Grace, K., Salvatier, J., Dafoe, A., Zhang, B., & Evans, O., (2018). When will AI exceed human performance? Evidence from AI experts. *Journal of Artificial Intelligence Research, 62*, 729–754.
2. Bagnell, J. A., Bradley, D., Silver, D., Sofman, B., & Stentz, A., (2010). Learning for autonomous navigation. *IEEE Robotics & Automation Magazine, 17*(2), 74–84.
3. Abbass, H. A., (2019). Social integration of artificial intelligence: Functions, automation allocation logic and human-autonomy trust. *Cognitive Computation, 11*(2), 159–171.
4. Bekir, E., (2007). *Introduction to Modern Navigation Systems*. World Scientific.
5. Watzenig, D., & Horn, M., (2017). Introduction to automated driving. In: *Automated Driving* (pp. 3–16). Springer, Cham.
6. Roshanianfard, A., Noguchi, N., Okamoto, H., & Ishii, K., (2020). A review of autonomous agricultural vehicles (The experience of Hokkaido University). *Journal of Terramechanics, 91*, 155–183.
7. Miranda, V. R., Rezende, A., Rocha, T. L., Azpúrua, H., Pimenta, L. C., & Freitas, G. M., (2022). Autonomous navigation system for a delivery drone. *Journal of Control, Automation and Electrical Systems, 33*(1), 141–155.
8. Joo, S. H., Manzoor, S., Rocha, Y. G., Bae, S. H., Lee, K. H., Kuc, T. Y., & Kim, M., (2020). Autonomous navigation framework for intelligent robots based on a semantic environment modeling. *Applied Sciences, 10*(9), 3219.
9. Polvara, R., Sharma, S., Wan, J., Manning, A., & Sutton, R., (2018). Obstacle avoidance approaches for autonomous navigation of unmanned surface vehicles. *The Journal of Navigation, 71*(1), 241–256.
10. Wang, Y., O'Keeffe, J., Qian, Q., & Boyle, D. E., (2022). *KinoJGM: A Framework for Efficient and Accurate Quadrotor Trajectory Generation and Tracking in Dynamic Environments.* arXiv preprint arXiv:2202.12419.
11. Pedersen, T. A., Glomsrud, J. A., Ruud, E. L., Simonsen, A., Sandrib, J., & Eriksen, B. O. H., (2020). Towards simulation-based verification of autonomous navigation systems. *Safety Science, 129*, 104799.
12. Mousazadeh, H., (2013). A technical review on navigation systems of agricultural autonomous off-road vehicles. *Journal of Terramechanics, 50*(3), 211–232.
13. Anavatti, S. G., Francis, S. L., & Garratt, M., (2015). Path-planning modules for autonomous vehicles: Current status and challenges. In: *2015 International Conference on Advanced Mechatronics, Intelligent Manufacture, and Industrial Automation (ICAMIMIA)* (pp. 205–214). IEEE.
14. Abdallaoui, S., Aglzim, E. H., Chaibet, A., & Kribèche, A., (2022). Thorough review analysis of safe control of autonomous vehicles: Path planning and navigation techniques. *Energies, 15*(4), 1358.
15. Alessandri, A., Donnarumma, S., Martelli, M., & Vignolo, S., (2019). Motion control for autonomous navigation in blue and narrow waters using switched controllers. *Journal of Marine Science and Engineering, 7*(6), 196.
16. Guo, H., Cao, D., Chen, H., Lv, C., Wang, H., & Yang, S., (2018). Vehicle dynamic state estimation: State of the art schemes and perspectives. *IEEE/CAA Journal of Automatica Sinica, 5*(2), 418–431.

17. Shit, R. C., (2020). Precise localization for achieving next-generation autonomous navigation: State-of-the-art, taxonomy and future prospects. *Computer Communications, 160*, 351–374.
18. Özaslan, T., Loianno, G., Keller, J., Taylor, C. J., Kumar, V., Wozencraft, J. M., & Hood, T., (2017). Autonomous navigation and mapping for inspection of penstocks and tunnels with MAVs. *IEEE Robotics and Automation Letters, 2*(3), 1740–1747.
19. Chatterjee, A., & Matsuno, F., (2007). A neuro-fuzzy assisted extended Kalman filter-based approach for simultaneous localization and mapping (SLAM) problems. *IEEE Transactions on Fuzzy Systems, 15*(5), 984–997.
20. Alatise, M. B., & Hancke, G. P., (2020). A review on challenges of autonomous mobile robot and sensor fusion methods. *IEEE Access, 8*, 39830–39846.
21. Wooden, D., Malchano, M., Blankespoor, K., Howardy, A., Rizzi, A. A., & Raibert, M., (2010). Autonomous navigation for BigDog. In: *2010 IEEE International Conference on Robotics and Automation* (pp. 4736–4741). IEEE.
22. Ananda, M. P., Bernstein, H., Cunningham, K. E., Feess, W. A., & Stroud, E. G., (1990). Global positioning system (GPS) autonomous navigation. In: *IEEE Symposium on Position Location and Navigation. A Decade of Excellence in the Navigation Sciences* (pp. 497–508). IEEE.
23. Haugh, B. A., Sparrow, D. A., & Tate, D. M. (2022). *Status of Test, Evaluation, Verification, and Validation (TEV&V) of Autonomous Systems*. Institute for Defense Analyses.
24. Liu, P., Yang, R., & Xu, Z., (2019). How safe is safe enough for self-driving vehicles? *Risk Analysis, 39*(2), 315–325.
25. Gu, Y., Goez, J. C., Guajardo, M., & Wallace, S. W., (2021). Autonomous vessels: State of the art and potential opportunities in logistics. *International Transactions in Operational Research, 28*(4), 1706–1739.
26. Singh, R., & Nagla, K. S., (2019). Comparative analysis of range sensors for the robust autonomous navigation: a review. *Sensor Review, 40*(1), 17–41.
27. Shalal, N., Low, T., McCarthy, C., & Hancock, N. (2013). A review of autonomous navigation systems in agricultural environments. *SEAg 2013: Innovative Agricultural Technologies for a Sustainable Future*.
28. Lee, S., Kwang-Joon, Y., & Jangmyung, L., (2012). *Frontiers of Intelligent Autonomous Systems*. Springer.
29. Katal, A., & Singh, N. (2022). Artificial neural network: models, applications, and challenges. *Innovative Trends in Computational Intelligence*, 235–257.
30. Singh, N., Prateek, M., & Chauhan, P., (2019). An efficient path planning algorithm for networked robots using modified optimization algorithm. In: *International Journal of Innovative Technology and Exploring Engineering* (Vol. 8, No. 12, pp. 2738–2743). Blue Eyes Intelligence Engineering and Sciences Engineering and Sciences Publication – BEIESP.
31. Le, A. V., Prabakaran, V., Sivanantham, V., & Mohan, R. E. (2018). Modified a-star algorithm for efficient coverage path planning in tetris inspired self-reconfigurable robot with integrated laser sensor. *Sensors, 18*(8), 2585.

CHAPTER 11

Autonomous Navigation: Present and Future

ARPITA JADHAV BHATT, ANKITA VERMA, and DEEPTI SINGH

Department of CSE&IT, Jaypee Institute of Information Technology, Noida, Uttar Pradesh, India

ABSTRACT

Autonomy is building its benchmark on becoming a formidable force. Autonomous systems can play a crucial role in assisting humans in a variety of real-time scenarios. This can potentially be in a broad range of applications, namely humanoid robots, driverless cars, military systems, carts in grocery stores, delivery robots, agricultural machines, etc. The autonomous navigation system requires no or minimal human intervention to choose its own path and navigate through its nearby areas. At present, these systems are also used to assist humans in problematic situations, i.e., situations where humans cannot go. For example, inside dangerous places, inspecting nuclear reactors to check for faults, inaccessible places such as aeroengines, etc. Currently, these systems are used in robotics, self-driving cars, military operations, agriculture, autonomous wheelchair systems, lawn mowers, surveillance systems, etc. Surveys and market analysis of these autonomous systems have predicted significant growth in the near future due to the increased demand for autonomous robots, cars, and drones in recent years. Research studies have predicted that autonomous navigation will become more natural in the coming years. Humans will see more autonomous vehicles like cars, which are already launched with improved connectivity and security. This chapter aims to

Advances in Autonomous Navigation Through Intelligent Technologies.
Niharika Singh, Thipendra P. Singh, Brian Azzopardi (Eds.)

provide the present applications of these systems and applications of these systems in the near future.

11.1 INTRODUCTION

The present era has brought a combination of 'Artificial Intelligence (AI)' and 'Autonomous Systems.' Autonomous systems are used in self-driving cars, drones, robots in warehouses, deep sea, software agents, and space exploration. Apart from that, chatbots in trading and websites are very prominent. AI is widely used in the form of machine learning and deep learning, and an increase in the availability of large datasets from different domains like image processing, detection of diseases, spam mail, IoT malware, etc., makes its integration easy with autonomous devices [1]. The convergence of these digital technologies is increasing quickly, making them very powerful. The autonomous systems are employed in various organizations, hospitals, public as well as private sectors, defense sectors, medical and industries, etc. The AI embedded in these systems can improve work conditions for humans and reduce strenuous tasks for them. These can be used for automatic heavy lifting of weights, packing of goods and food products in industries, battlegrounds, drilling works, etc. These systems are capable of working without human or minimal human intervention and can be controlled from outside or from a workstation. These smart systems can conduct dialogs with clients or customers in online call centers; help in buying and selling stocks in large quantities in a fraction of seconds; direct cars to carry passengers; and prevent collisions in case of emergency. From past research studies, a plethora of applications of these systems in several domains such as agriculture, defense, e-commerce, transportation, smart healthcare, etc., are proposed [2–4]. This chapter discusses in detail the different sectors in which autonomous systems are deployed, their present implementation, usage, and future perspectives.

11.2 GROWTH OF THE AUTONOMOUS MARKET

Autonomous navigation systems are becoming ubiquitous in diverse industrial, agricultural, and surveillance activities as well as gaining popularity in household activities. With the advent of novel sensing technologies, the development of robust artificial and machine learning algorithms, and the support of complex infrastructure, the commercial applicability of these smart systems is increasing in leaps and bounds. The market for autonomous navigation is valued at USD 13.89 billion, and it is expected to grow at a Compound Annual Growth Rate (CAGR) of 16.59% between 2020 and 2030 [42]. From the perspective of different platforms on which autonomous navigation systems are operating, they can be widely categorized as Unmanned Ground Vehicles (UGVs), Unmanned Aerial Vehicles (UAVs), and Unmanned Surface Vehicles (USVs). UGVs operate at the ground level navigating their ways through different landscapes avoiding obstacles [5, 6]. The marketplace for UGVs is projected to grow at a rate (CAGR) of 5.7% from 2022 to 2027, i.e., from USD 2.7 billion in 2022 to USD 3.6 billion in 2027 [43]. UAVs (aka drones) operate at an aerial level and find their wider applicability as surveillance drones and delivery drones [7]. UAVs market has a projected CAGR of 16.4% from 2021 to 2026 with market share rising from USD 27.4 billion to USD 58.4 billion [44]. Unmanned Surface Vehicles (USVs) are expected to assist in various naval activities reducing the cost and time of operations [8]. During the forecasted period, the unmanned surface vehicles market is expected to increase at a CAGR of 11.0%, i.e., from USD 616 million in 2021 to USD 1,038 million in 2026 [45]. Thus, from the market analysis, it is evident that research and development activities in this domain need to be further accelerated to aid the growing market.

11.3 APPLICATIONS OF AUTONOMOUS SYSTEMS WITH THEIR FUTURE IMPLICATIONS IN VARIOUS INDUSTRIES

11.3.1 TRANSPORT

In recent years, there has been an upsurge in the need for autonomy to achieve reliable, swift, and safe transportation. To this end, autonomous navigation systems developed using the process of SLAM (simultaneous localization and mapping) are capable of automatically creating dynamic maps of the environment, navigating the device by planning suitable paths, and avoiding all obstacles encountered along the way [9]. From human-guided partial automation, research in this domain is rapidly progressing towards unmanned full automation. However, to facilitate this, we require accurate passive and active sensors, efficient techniques of simultaneous localization and mapping, and robust steering mechanisms based on artificial intelligence and machine learning algorithms to navigate the robot safely while avoiding obstacles. Some of the application areas of mobile robots in transport are discussed below:

1. **Self-Driving Vehicles (SDVs):** These are one of the most significant applications of autonomous navigation systems. The primary motivation behind this technological advancement is to reduce mishaps caused by human errors. In addition, the use of SDVs will lead to better utilization of road capacity, less traffic congestion, increased fuel efficiency, and enhanced mobility for non-drivers such as disabled individuals and the elderly [10, 40, 41].
2. **Transport Robots in Industry:** As per Dr. Khasha Ghaffarzadeh, Research Director at IDTechEx, mobile robots in logistics and material transportation will grow to be a market of 75 billion dollars in 2027. The use of mobile robots for simple repetitive activities like transporting materials in warehouses, automatic packaging, dispatching, stock maintenance, etc., increases the efficiency and swiftness of the supply chain. Some of the practical examples are Jungheinrich APM (auto pallet mover), SRX1 and SRX3, KMR iiwa, etc.
3. **Delivery Drones:** Drone delivery systems are garnering great interest from industry and academia as they facilitate swift, safe, and cost-effective delivery of goods. UPS Flight Forward Inc. is one of the initial companies targeting to develop a network for drone deliveries in the USA [11]. In a similar line of work, Wing (a division of Alphabet) aims to develop drones for freight delivery [12].

11.3.1.1 FUTURE PERSPECTIVES IN TRANSPORTATION

- For SDVs, safety is one of the biggest concerns. It becomes important to determine the acceptable level of risks in SDVs. We need to verify whether these vehicles are designed to reduce the chances of accidents or pose new threats like accidental system failure, cyber-attacks, breach of privacy, etc. It is evident that extensive research needs to be conducted to verify all the security and safety concerns in order to increase the market presence and social acceptability of SDVs.

- For autonomous navigation systems to flourish in the domain of transportation, it requires great support from infrastructure. It is highly critical to focus on the development of cost-effective technological solutions which promote optimal usage of resources.
- Significant technical improvements are required to reach level 5 of full automation. Currently, Level 4 of autonomy is achieved which ensures reliable performance in a controlled environment.
- Regulatory approvals and robust testing are required for their commercial reach.

11.3.2 AGRICULTURE

Autonomous navigation systems find their wide applicability in various tasks related to agriculture and farming such as cultivation, planting, fertilizer spraying, harvesting, etc. The use of an autonomous robotic system may assist farmers in performing hazardous tasks, save manual labor and time, and provide them with crucial insights for effective decision-making [13]. They are the need of the hour for meeting the increasing food requirement of the world, utilizing the environmental resources effectively with precision farming, and overcoming the scarcity of manual labor for farming [14]. For the development of an efficient robotic navigation system for agricultural requirements, various factors such as weather conditions, landscape variations, types of vegetation, mobility of obstacles, etc., have to be taken into consideration. Owing to this, the researchers are actively working in this domain and developing competent systems for sensing, localization, mapping, path scheduling, and hindrance avoidance [4]. Autonomous navigation is a crucial part of any robotic system as it enables planned and safe movement of the robots in its environment. For planning the sequence of movements of the robot and avoiding obstacles, navigation sensors need to collect relevant information such as the current position of the robot, its speed, the direction of the movement, and the knowledge of the environment it's moving in. Some of the primary areas of farming which are benefited from autonomous navigation systems are mentioned below:

1. **Autonomous Seeders and Planters:** Automatic seeders are designed to automate the process of sowing seeds in a regular pattern in the fields. These robots are equipped to assess the land parameters and plant seeds with accurate seed spacing for better utilization of resources like land, water, sun, etc. [15, 16].
2. **Autonomous Harvesting:** It is the ideal process for automation as it is conducted several times a year and has to be done with precision to avoid any damage to the crop [17]. Since the advent of the autonomous system, they have been utilized for harvesting various crops such as cucumber [18], strawberries [19], apples [20], sweet pepper [21], etc.
3. **Crop Row Navigation for Inspection:** An autonomous robot needs to collect various insights about the crops in order to identify contaminated plants and crops, check for the presence of fungus or molds, and search the fields for any intruders (animals or pests). To this end, the autonomous robot needs to correctly locate the beginning and end of the rows and perform several turning maneuvers successfully [22].
4. **Autonomous Crop Treatment:** Weed identification and removal is one of the significant crop treatment tasks garnering the benefits of automation [23]. The use of computer autonomous navigation system for spraying pesticides, herbicides, and fertilizers, etc., promotes precise coverage of affected crop yields in a time-efficient manner [24].
5. **Miscellaneous:** Autonomous navigation systems are also assisting in various tasks such as tree counting, management of livestock, classification of plant species, determination of crop maturity, pruning, grafting, etc. [14].

11.3.2.1 FUTURE PERSPECTIVES IN AGRICULTURE

- Various sensing devices such as global positioning systems, RFID, and laser range sensors are usually used for perceiving the environment. From the very beginning, GPS-based systems such as Real-Time Kinematic Global Positioning System (RTK-GPS) and

Differential Global Positioning System (DGPS), etc., are extensively used as navigation sensors [26]. Vision-based sensing is a cost-effective and wholesome way to perceive the environment of the robot. Beginning with 1D vision systems [27], developing successful 2D systems [28], and further improving with 3D vision systems [29], vision-based sensing is adapting to more complex applications. Laser-based scanners are also widely utilized for agricultural applications as they are well adapted to different weather situations and terrains and are capable of providing a robust range of data. However, gathering relevant perception from the environment is dependent on the efficacy of the sensors used and the decision of which sensor or the combination of sensors would be apt in the given scenario is crucial.

- One size fits all design of guided navigation system is not possible for the agricultural needs owing to different vegetations, terrains, landscapes, and weather conditions. Thus, the dedicated harvesting, seeding, and other activities robots are to be developed specifically for the targeted crop.
- In order to truly harness the potential of automation in agriculture, dedicated infrastructure such as complex robot mechanisms, networking resources, human-computer interfaces, and a technologically sound workforce is required. Facilitating these is far from a feasible task for developing and underdeveloped countries [30].

11.3.3 DEFENSE

The advancements in robotics along with the development of autonomous navigation systems have transformed the defense and military sector globally with smart functionalities. At present, various robots are being employed in ground missions to protect a nation from future attacks [31]. The automated robots help in augmenting the human capabilities when they are responding to possible threats without harming the human soldiers. For example, the robots are capable of neutralizing the bombs in the war zone. At present, Governments have already started investing in robotics for accelerating military tasks. The military sector leverages robotics in order to provide their soldiers a benefit at the base level. The integration of artificial intelligence provides automation to the robots and is still controlled by the warriors at the battleground to prevent autonomous random killing. Autonomous navigation systems are widely used in the defense sector and have been listed below [31]:

1. **Wearable Robots:** These robots are helping the fighters to pack their loads efficiently. The combination of AI with robotics helps in carrying a bulk of supplies from one location to another. These help the human soldiers by reducing their workload and allowing them to focus their attention on other crucial areas.
2. **Mobile Robots:** These play a wide range of roles in conducting military operations from investigation to neutralizing bombs. They are equipped with high-speed cameras and high-tech sensors in order to carry out several missions efficiently. They are also equipped with robotic arms for neutralizing bombs and suspicious objects without harming soldiers. They can work on tracks, wheels, and battery with good communication.
3. **Surveillance Robots:** These are primarily used in the defense and military sector for the betterment of the citizens. The ground-based defense systems use these robots with weapons as well as high-sensor cameras which are equipped with night vision and infrared to monitor either buildings or warzones without sending human soldiers. This minimizes fatalities.
4. **Combat Robots:** These robots are widely used in the military sector as they have advanced functionalities, namely fire support, managing damages in battle, laying mines, aerial refueling, etc. They are equipped with automating features in order to minimize collateral damage effectively and efficiently on the battlegrounds.
5. **Swarm of Drones:** These drones are specially designed for powerful observation in war zones. They are surveillance drones that are small in size but are equipped with high-battery life. They also have high-tech cameras and infrared detection capabilities.

11.3.3.1 FUTURE PERSPECTIVES IN DEFENSE AND MILITARY

- The high-tech cameras can be equipped with infrared detection capabilities to provide a high-quality aerial view, which is very essential in the defense sector for observing activities in battle zones as well as detecting potential threats to combat.

- In the future, voice-controlled robotic vehicles can be used, which rely on voice instructions from a speech recognition module in order to conduct essential activities. These can be equipped with speeches and push buttons with circuits to control movements.

11.3.4 HEALTHCARE

Autonomous systems, along with robotics, are bringing innovations to the healthcare sector. Artificial Intelligence (AI) is an emergent component in the autonomous systems in the healthcare domain and is expected to lead to a global market value of 34 billion dollars by the year 2025 [32]. The technologies that employ AI [25] and machine learning (ML) approaches offer clinics and hospitals the capability to gain new efficiencies and speed up their potential in order to provide effective healthcare to their patients [33]. At present, Wipro is providing AI and ML-based platforms to healthcare organizations and research labs in order to improve their approaches for the discovery of drugs, genetics research, and diagnosing the medical conditions of their patients. Healthcare organizations are moving towards autonomous robotic operations and benefiting from the self-driving robots that are capable of handling drug deliveries and COVID-19-related sanitization tasks [33].

Autonomous systems are widely used in healthcare, such as autonomous surgical robots (da Vinci Surgical System). Automated insulin delivery systems, which can autonomously regulate the insulin level of the human body, are used. Systems that consist of infusion pumps and sensors that infuse the correct dose of a painkiller are widely used in healthcare [34]. These automated systems have resulted in many advantages, such as improved diagnosis of patients' health conditions, safer operational environments, and more focus on essential tasks for the doctors. With the help of self-driving robots, surgeons can focus on critical tasks. These systems also promote contactless operations. Currently, many autonomous systems are being used, and a few of them have been described below:

1. **Da Vinci Surgical System:** It is an automatic surgical system that uses a minimally invasive surgical approach. The system was manufactured by Intuitive Surgical [35]. It is widely used

for treating prostatectomies, cardiac valve repair, and for renal as well as gynecologic surgeries. Since 2012, it has been used in approximately 2,00,000 surgeries, mainly for prostate removal and hysterectomies.

2. **Automated Insulin Delivery System:** These systems can be automated or semi-automated and are designed to assist medical patients with type-1 diabetes [35]. These systems automatically adjust the insulin delivery to the patients and help them control their blood sugar levels.

   At present, these systems are still in the development stages. For these systems to be trusted in the sensitive context of healthcare, policymakers have to consider a wide range of issues such as the safety of patients, information privacy, security of medical records, precision during surgeries, etc.

11.3.4.1 FUTURE PERSPECTIVES IN HEALTHCARE

- The public satisfaction of autonomous systems in the healthcare domain appears to be comparatively varied and unclear. Policymakers have to work on developing systems that are highly precise.
- At present, the available systems (till October 2020) can only deliver a single hormone, insulin. Other systems in the developing stage can aim to improve the existing systems by adding more hormones that can be delivered according to the patient's health condition.
- It is also predicted that these systems might assist the patients in their homes.

11.3.5 E-COMMERCE

Electronic commerce (E-commerce) employs buying and selling of goods as well as services, or it involves transmissions of data or funds over interconnected networks, primarily the internet. The business transactions can occur as business-to-consumer (B2C), consumer-to-consumer,

consumer-to-business, or business-to-business (B2B). The last decade has seen widespread use of e-commerce applications such as Amazon, Flipkart, and eBay globally [37]. Moreover, after the COVID-19 pandemic, demands for food ordering and delivery applications like Zomato and Swiggy have also increased. These e-commerce applications have contributed to substantial growth in online retailing. Additionally, with the increased use of smartphones, Mobile e-commerce (M-commerce) is on the rise. It features online sales of products using mobile devices such as smartphones or tablets, including mobile banking, shopping, payments, etc. Currently, mobile chatbots are heavily used to provide e-commerce opportunities to organizations and enterprises by allowing their consumers to complete transactions with the companies through text or voice conversations. An E-commerce application requires managing the infrastructure globally with a large network of amenities, professionals, and machines. Nowadays, the use of autonomous robots and material handling systems is providing great efficiencies to these organizations. The autonomous systems are designed to take over redundant and risky tasks that were conventionally delegated to human employees [37]. E-commerce applications are now using autonomous robots for e-commerce fulfillment and at distribution centers to meet the increasing demands of their customers.

11.3.5.1 AUTONOMOUS SHOPPING SYSTEMS

11.3.5.1.1 State of the Art

These systems are built on the technology in which the consumers or customers can delegate important parts of the shopping process, which may include shopping decisions and tasks [38]. This system automatically reaches a series of conclusions for their customers, for instance, which and how many items to buy based on past data or input data. At present, many such autonomous systems are used. For instance, Boxed's Concierge automatically places orders on behalf of its clients when it forecasts that they are running short or low on a product or item. It is based on predictive analysis. Another popularly used autonomous system is Samsung's 'Family Hub' refrigerator, which automatically places orders for groceries by scanning the items in the refrigerator. In the near future, more advanced autonomous systems will be capable of taking complex decisions like humans. For example, the systems may choose apparel for a client based on his/her past choices. These systems are designed to achieve a definite set of largely deterministic steps and are built using simple reflex-like rules. These systems learn, develop, and are capable of changing their functionalities based on the input of operational as well as contextual information.

Moreover, virtual shopping assistants that provide personalized recommendations based on artificial algorithms are heavily used these days [38]. These systems identify and match the products and services of the customers according to their requirements. The recommender system applies a machine learning algorithm and offers a supporting role in the customer shopping process, helping them make decisions. With advancements in the area of Artificial Intelligence (AI), recommender systems are rapidly transforming into autonomous shopping systems.

11.3.5.2 AUTONOMOUS SHOPPING CART

11.3.5.2.1 State of the Art

Several types of research in robotics have been conducted to date to develop autonomous navigation systems like robots that can assist humans in their everyday chores. These are called service robots that can assist humans in supermarkets. These systems are designed to help persons with disabilities [39]. They can use these autonomous robots for carrying their groceries while they are shopping. These systems are even capable of computing the shortest tour for a robot by using the Traveling Salesman Problem (TSP), such that it can traverse all aisles required by a customer based on their shopping list. Additionally, there have been several autonomous robots that are capable of performing tasks in the human environment such as rearranging simple objects, indoor navigation, speech recognition systems, etc. [39].

11.3.5.3 Future Perspectives in E-Commerce Applications

- As technologies are becoming highly autonomous and require no or minimal human intervention, the policymakers of e-commerce applications must ensure that the future human-machine interactions are more focused on their clients' needs and preferences.

- Autonomous systems must be designed with the ability to maintain and repair relationships with human collaborators.
- In the future, recommender systems must be equipped with monthly budget scanning and monitoring of the budget of their users so that they do not indulge in impulsive shopping.
- At present, many autonomous shopping carts are employing the TSP algorithm to find the shortest path. Future works must consider the application of nature-inspired algorithms such as particle swarm optimization (PSO), ant colony optimization (ACO), Tabu search, and simulated annealing to solve TSP and generate tours.
- Other algorithms, namely, rapidly-exploring random trees, can also be employed for creating the path matrix in the systems to reduce the computational costs for finding the optimal paths.

11.3.6 INDUSTRIES AND WAREHOUSES

In the present era, the shipping industry, warehouses, and large and small-scale industries are employing autonomous systems for manufacturing and delivering their products. These autonomous navigation systems are widely used in industries and warehouses for material handling, picking, and sorting industrial materials, heavy lifting, automated packing, transportation, and delivery of goods [37].

11.3.6.1 STATE OF THE ART

In material handling (MH), the robots generally have an arm tool called a 'gripper.' It can be at a fixed place or can be mobile. These robots can also be equipped with hot metal detectors in granulated or mill environments. These robots are very useful in factories and warehouses that deal with various hazardous materials because they can eliminate the need for human involvement. These robots also contain technologies for tracking conveyors as well as detecting collisions, which are very important in providing safety features in monorail systems and warehouses.

Moreover, in industries and food factories, picking, and sorting goods or products are the most labor-intensive activities. It requires a lot of hand-eye coordination and dedication from the workers. Humans who occupy these industrial roles may have to walk or travel several miles to fulfill their roles and responsibilities. Autonomous picking robots can assist in sorting and packing goods so that humans do not have to. These robots can be either free-roaming or can be attached to conveyors. These systems can even be linked with the 2-dimensional or 3-dimensional sensors or can be equipped with scanners to recognize the objects as well as sort them in correct dimensions [37].

At present, in industries and warehouses, autonomous robots are used for heavy lifting. These systems are capable of lifting thousands of pounds. They are equipped with a safety laser scanning technique with a 360° view of nearby surroundings. These systems can move steadily with safety measures across the warehouse floor.

After material handling, heavy lifting, and picking and sorting of goods, next comes automating the process of packaging the goods. These days, autonomous systems are employed for packaging the goods with minimal human intervention. These systems are equipped with a robotic arm and are integrated with sensors and cameras to deduce minute specifications. The systems can pack goods faster than human beings and also offset the physical risks involved in packing. These packing robots may include robotic case erectors, sealers, and several other tools that assist in the packaging of goods. At present, these types of robots are used in food industries and are employed to pack even fragile food products like eggs. Apart from that, these robots are widely used for transporting goods and products from one section of the industry or warehouse to another section without any human intervention. These autonomous transportation robots can be equipped with 2-dimensional and 3-dimensional light detecting and ranging (LiDAR) sensors for navigating across the warehouse. These systems are globally used in warehouses [37].

11.3.6.2 FUTURE PERSPECTIVES IN INDUSTRIES AND WAREHOUSES

- Autonomous navigation systems involving physical robots will transform the warehousing process, while virtual assistants will operate as personal digital assistants that interpret customers' needs and make decisions on their behalf.

- These systems can offer significant improvements in production and efficiency and improve customer satisfaction.
- The autonomous systems can enhance the revenues of industries by improving delivery speed and perfect order accomplishment rates, leading to improved customer satisfaction.

11.4 CONCLUSION

We've attempted to highlight some of the several application domains of autonomous navigation systems in this chapter. As demonstrated by current market trends, the applicability of these smart systems will ramp up, with a profound influence on industries, military, agriculture, and transportation sectors. To aid this, a great deal of emphasis should be devoted to the development of reliable, cost-effective, and sustainable technologies.

KEYWORDS

- **autonomous navigation**
- **autonomous systems applications**
- **autonomous vehicles**
- **collisions**
- **delivery robots**
- **light detecting and ranging**
- **sustainable technologies**

REFERENCES

1. Zhilenkov, A. A., & Epifantsev, I. R., (2018). Problems of a trajectory planning in autonomous navigation systems based on technical vision and AI. In: *2018 IEEE Conference of Russian Young Researchers in Electrical and Electronic Engineering (EIConRus)* (pp. 1032–1035). IEEE.
2. Mohamed, S. A., Haghbayan, M. H., Westerlund, T., Heikkonen, J., Tenhunen, H., &Plosila, J., (2019). A survey on odometry for autonomous navigation systems. *IEEE Access, 7*, 97466–97486.

3. Chao, Q., Huikun, B., Weizi, L., Tianlu, M., Zhaoqi, W., Ming, C. L., & Zhigang, D., (2020). A survey on visual traffic simulation: Models, evaluations, and applications in autonomous driving. In: *Computer Graphics Forum* (Vol. 39, no. 1, pp. 287–308).
4. Shalal, N., Low, T., McCarthy, C., & Hancock, N. (2013). A review of autonomous navigation systems in agricultural environments. *SEAg 2013: Innovative Agricultural Technologies for a Sustainable Future*.
5. Brust, M. R., & Strimbu, B. M., (2015). A networked swarm model for UAV deployment in the assessment of forest environments. In: *2015 IEEE Tenth International Conference on Intelligent Sensors, Sensor Networks and Information Processing (ISSNIP)* (pp. 1–6). IEEE.
6. Stolfi, D. H., Brust, M. R., Danoy, G., & Bouvry, P., (2020). Emerging inter-swarm collaboration for surveillance using pheromones and evolutionary techniques. *Sensors*, *20*(9), 2566.
7. McNeal, G. S., (2016). Drones and the future of aerial surveillance. *Geo. Wash. L. Rev.*, *84*, 354.
8. Costanzi, R., Fenucci, D., Manzari, V., Micheli, M., Morlando, L., Terracciano, D., & Tesei, A., (2020). Interoperability among unmanned maritime vehicles: Review and first in-field experimentation. *Frontiers in Robotics and AI*, *7*, 91.
9. Bailey, T., & Durrant-Whyte, H., (2006). Simultaneous localization and mapping (SLAM): Part II. *IEEE Robotics & Automation Magazine*, *13*(3), 108–117.
10. Litman, T., (2020). *Autonomous Vehicle Implementation Predictions: Implications for Transport Planning*.
11. Schneider, D., (2020). The delivery drones are coming. *IEEE Spectrum*, *57*(1), 28–29. https://doi.org/10.1109/MSPEC.2020.8946304.
12. Miranda, V. R., Rezende, A., Rocha, T. L., Azpúrua, H., Pimenta, L. C., & Freitas, G. M., (2022). Autonomous navigation system for a delivery drone. *Journal of Control, Automation and Electrical Systems*, *33*(1), 141–155.
13. Grift, T., Zhang, Q., Kondo, N., & Ting, K. C., (2008). A review of automation and robotics for the bio-industry. *Journal of Biomechatronics Engineering*, *1*(1), 37–54.
14. Bergerman, M., Van, H. E., Billingsley, J., Reid, J., & Mingcong, D., (2013). IEEE robotics and automation society technical committee on agricultural robotics and automation. *IEEE Robotics & Automation Magazine*, *20*(2), 20–23.
15. Litvinov, M. A., Moskovskiy, M. N., & Smirnov, I. G., (2018). Automatic system of intelligent seed rate control for selection seeders. In: *2018 IEEE East-West Design & Test Symposium (EWDTS)* (pp. 1–5). IEEE.
16. Pawar, M. S., Ugale, V. D., Dawada, H., Deore, P., & Deore, U., (2020). Automatic seedling robot. *International Journal of Research and Analytical Reviews (IJRAR)*. E-ISSN, 2348-1269.
17. Bac, C. W., Van, H. E. J., Hemming, J., & Edan, Y., (2014). Harvesting robots for high-value crops: State-of-the-art review and challenges ahead. *Journal of Field Robotics*, *31*(6), 888–911.
18. Van, H. E. J., Hemming, J., Van, T. B. A. J., Kornet, J. G., Meuleman, J., Bontsema, J., & Van, O. E. A., (2002). An autonomous robot for harvesting cucumbers in greenhouses. *Autonomous Robots*, *13*(3), 241–258.

19. Xiong, Y., Ge, Y., Grimstad, L., & From, P. J., (2020). An autonomous strawberry-harvesting robot: Design, development, integration, and field evaluation. *Journal of Field Robotics*, *37*(2), 202–224.
20. Baeten, J., Donné, K., Boedrij, S., Beckers, W., & Claesen, E., (2008). Autonomous fruit picking machine: A robotic apple harvester. In: *Field and Service Robotics* (pp. 531–539). Springer, Berlin, Heidelberg.
21. Lehnert, C., English, A., McCool, C., Tow, A. W., & Perez, T., (2017). Autonomous sweet pepper harvesting for protected cropping systems. *IEEE Robotics and Automation Letters*, *2*(2), 872–879.
22. Bonadies, S., & Gadsden, S. A., (2019). An overview of autonomous crop row navigation strategies for unmanned ground vehicles. *Engineering in Agriculture, Environment and Food*, *12*(1), 24–31.
23. Kesavan, K., Monish, R., & Murali, S., (2018). Automatic weed removal system. *Int. J. Res. Eng. Sci. Manag.*
24. Hamner, B., Singh, S., & Bergerman, M., (2010). Improving orchard efficiency with autonomous utility vehicles. In: *2010 Pittsburgh, Pennsylvania* (p. 1). American Society of Agricultural and Biological Engineers.
25. Le, A. V., Prabakaran, V., Sivanantham, V., & Mohan, R. E. (2018). Modified a-star algorithm for efficient coverage path planning in Tetris inspired self-reconfigurable robot with integrated laser sensor. *Sensors*, *18*(8), 2585.
26. Eaton, R., Katupitiya, J., Siew, K. W., & Howarth, B., (2010). Autonomous farming: Modeling and control of agricultural machinery in a unified framework. *International Journal of Intelligent Systems Technologies and Applications*, *8*(1–4), 444–457.
27. Jørgensen, R. N., (2002). *Study on Line Imaging Spectroscopy as a Tool for Nitrogen Diagnostics in Precision Farming.* Risø, Denmark: Risø Natl. Lab.
28. Zhang, J., Kantor, G., Bergerman, M., & Singh, S., (2012). Monocular visual navigation of an autonomous vehicle in natural scene corridor-like environments. In: *2012 IEEE/RSJ International Conference on Intelligent Robots and Systems* (pp. 3659–3666). IEEE.
29. Vázquez-Arellano, M., Griepentrog, H. W., Reiser, D., & Paraforos, D. S., (2016). 3-D imaging systems for agricultural applications—A review. *Sensors*, *16*(5), 618.
30. Hajjaj, S. S. H., & Sahari, K. S. M., (2016). Review of agriculture robotics: Practicality and feasibility. In: *2016 IEEE International Symposium on Robotics and Intelligent Sensors (IRIS)* (pp. 194–198). IEEE.
31. Analytics Insight, *How is the Military and Defense Sector Utilizing Robotics*? [Online]. Available: https://www.analyticsinsight.net/how-is-the-military-and-defence-sector-utilizing-robotics/ (accessed on 27 April 2024).
32. Saracco, R. *Autonomous Systems in Healthcare*. [Online]. Available: https://cmte.ieee.org/futuredirections/2019/07/21/autonomous-systems-in-healthcare/ (accessed on 27 April 2024).
33. Wipro Limited, *Robotics Solutions for Healthcare Providers Technologies Key Takeaways*.
34. Johner-Institute, *Autonomous Systems: 4 Advantages for Medical Device Manufacturers*. [Online]. Available: https://www.johner-institute.com/articles/product-development/and-more/autonomous-systems (accessed on 27 April 2024).

35. Wikipedia, *Automated Insulin Delivery System.* [Online]. Available: https://en.wikipedia.org/wiki/Automated_insulin_delivery_system (accessed on 27 April 2024).
36. Chai, W., *What is E-Commerce?* [Online]. Available: https://www.techtarget.com/searchcio/definition/e-commerce (accessed on 27 April 2024).
37. Hokuyo, *How Autonomous Robots are Used in eCommerce.* [Online]. Available: https://hokuyo-usa.com/resources/blog/5-ways-autonomous-robots-are-utilized-manage-increased-demand-ecommerce (accessed on 27 April 2024).
38. De Bellis, E., &. Venkataramani, J. G., (2020). Autonomous shopping systems: Identifying and overcoming barriers to consumer adoption. *Journal of Retailing, 96*(1), 74–87.
39. Alves, R., Linhares, B. A., & Souza, J. R., (2018). Autonomous shopping cart: A new concept of service robot for assisting customers. *Proceedings – 15th Latin American Robotics Symposium, 6th Brazilian Robotics Symposium and 9th Workshop on Robotics in Education, LARS/SBR/WRE 2018* (Vol. 128, pp. 457–463).
40. Katal, A., & Singh, N., (2022). Artificial neural network: Models, applications, and challenges. In: Tomar, R., Hina, M. D., Zitouni, R., & Ramdane-Cherif, A., (eds.), *Innovative Trends in Computational Intelligence.* EAI/Springer Innovations in Communication and Computing. Springer, Cham. https://doi.org/10.1007/978-3-030-78284-9_11.
41. Singh, N., Prateek, M., & Chauhan, P., (2019). An efficient path planning algorithm for networked robots using modified optimization algorithm. In: *International Journal of Innovative Technology and Exploring Engineering* (Vol. 8, No. 12, pp. 2738–2743). Blue Eyes Intelligence Engineering and Sciences Engineering and Sciences Publication – BEIESP.
42. https://www.marketresearchfuture.com/reports/autonomous-navigation-market-7650.
43. https://www.marketsandmarkets.com/Market-Reports/unmanned-ground-vehicles-market-72041795.html.
44. https://www.marketsandmarkets.com/Market-Reports/unmanned-aerial-vehicles-uav-market-662.html.
45. https://www.marketsandmarkets.com/Market-Reports/unmanned-surface-vehicle-market-220162588.html

CHAPTER 12

Algorithm Analysis of Automatic Robots

JAS KARMAN SAGGAR[1] and NIHARIKA SINGH[2]

[1]*School of Computer Science, University of Petroleum and Energy Studies, Uttarakhand, India*

[2]*Lead Data Engineer, MPAC, Canada*

ABSTRACT

The quest for autonomous mobile robots capable of navigating unknown environments sans human intervention has spurred extensive research and development in recent times. These robots are categorized into manual, remote-controlled, and autonomous systems. Manual robots require significant human involvement, while remote-controlled ones are operated externally. Autonomous robots, the focus of recent advancements, make decisions independently based on environmental perception. Autonomous robots are further subdivided into programmable, non-programmable, adaptive, and intelligent categories. Programmable robots feature configurable control systems and are versatile but lack adaptability. Non-programmable robots are simpler, typically used in precision-demanding tasks. Adaptive robots autonomously adjust within certain parameters, while intelligent robots exhibit sophisticated cognitive capabilities, leveraging AI and machine learning for complex tasks. Path planning algorithms are pivotal for autonomous navigation. Traditionally, robots were tele-operated, but self-learning algorithms like reinforcement learning have broadened their autonomy. Algorithms like Dijkstra, A*, and genetic algorithms aid path planning, with 3D imaging enhancing navigation accuracy. Future advancements include reinforcement learning and optimization techniques like cuckoo search, paving the way for more efficient autonomous navigation in dynamic environments.

Advances in Autonomous Navigation Through Intelligent Technologies.
Niharika Singh, Thipendra P. Singh, Brian Azzopardi (Eds.)

12.1 INTRODUCTION

The autonomous mobile robot should be capable of navigating in an unknown environment without human interaction. In order to successfully reach its target while avoiding obstacles, proper path planning is crucial for mobile robots. In recent times, there has been a significant amount of research and development in the field of autonomous navigation in unknown environments [1]. The study shows that the robots can be divided into three categories: manually controlled robotic systems, remote-controlled robotic systems, and autonomous controlled robotic systems. Automated systems that require extensive human involvement for their functioning are termed as manual robots. These automatic systems need exceptional human authority for controlling various processes incorporated with the machines. Manual robots embrace a series of robotic systems, from elementary to highly advanced, all dealing with a specific control system according to their diligence. On the other side, a remote-controlled robot is defined as any vehicle teleoperated by a means that does not restrict its motion with an origin external to the device. In addition, the last and foremost category of robots is autonomous. Human-like robots that have the power to get to their conclusions and accomplish an action consequently are called autonomous robots. These robotic systems operate mainly by perceiving their surroundings, analyzing that information, and then taking action within that environment. When it comes to movement, their decision-making processes involve basic actions such as starting, stopping, and navigating around obstacles. However, they are not limited to these actions. This category of robots is further subdivided into:

- Programmable;
- Nonprogrammable;
- Adaptive;
- Intelligent.

12.1.1 PROGRAMMABLE AUTOMATIC ROBOT

A robot having actuator capabilities on each joint is a programmable robot. Depending on the type of use that is permitted, the robots may be programmable. After the robot is adjusted once to carry out power in the provided example and fixed arrangement, the capacity and utilization of the robots can be changed by reinventing them. The main drawback of this autonomous robot is that it can no longer be altered. It continues working whether or not its assignment has to be changed (in case of a crisis). These robots can be used for a variety of tasks, including spaceflight applications, adaptable mechanical autonomy, and industrial controls [2] (Figure 12.1).

FIGURE 12.1 Example of programmable robot.
Source: ROBOTIS. https://www.robotis.us/robotis-premium/

12.1.2 NON-PROGRAMMABLE ROBOT

One of the most fundamental types of robots is a non-programmable robot. This robot lacks a configurable control system, making it an exploiter rather than a robot. These types of robots have mechanical arms. These types of robots find applications in some devices including path guiders and medical products' carriers, and also some line follower robots. Non-programmable robots are often used in environments where a high degree of precision is required, such as in manufacturing or medical settings. They are also typically less expensive than programmable robots, making them a more accessible option for smaller businesses or organizations [2] (Figure 12.2).

FIGURE 12.2 Example of non-programmable robot.

Source: FANUC. https://www.fanuc.eu/il/en/robots

12.1.2.1 ADAPTIVE ROBOT

Adaptive robots are industrial robots that have the ability to autonomously adapt to a variety of process ranges. But compared to programmable robots, these robots are more sophisticated. They are capable of modification up to a certain point, and after review, they are capable of performing the required action in the modified area.

Most of these robots have sensors and control systems. Adaptive robots are mostly employed in welding and spraying systems. Many sorts of robots, including those with sensors and microprocessors for storing and processing data, are available, but intelligent robots are, as their name suggests, the most intelligent. Due to their capabilities for situation-based analysis and task execution, these robots' function with extremely high efficiency.

Robots with intelligence can detect pain, smell, taste, vision, and hearing. They can execute activities and expressions such as emotions, thoughts, and learning accordingly. Medical, military, and household appliance control systems all use these robots in some capacity. Additionally, with advancements in artificial intelligence and machine learning, these robots are becoming more sophisticated and able to adapt to new situations on their own. This has led to increased interest in their use in industries such as manufacturing, transportation, and agriculture [2] (Figure 12.3).

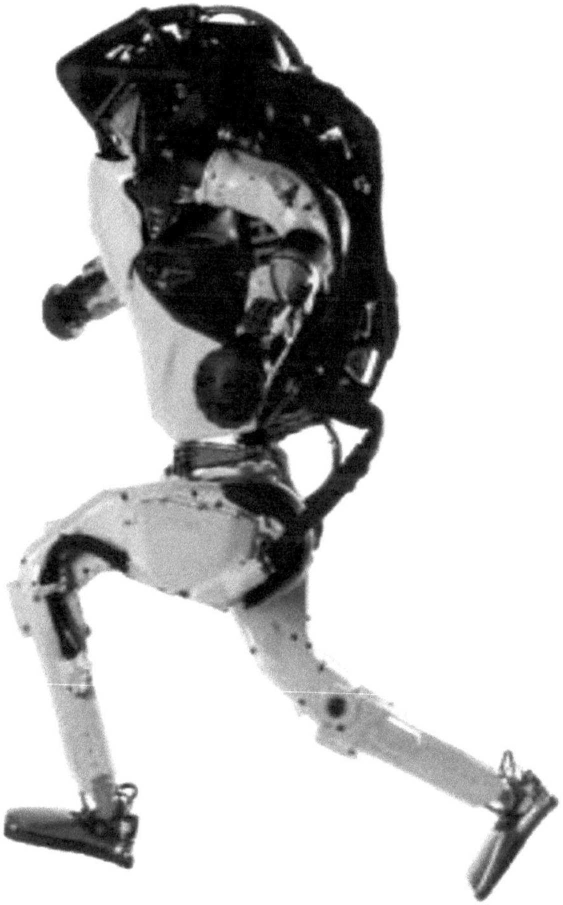

FIGURE 12.3 Example of autonomous robot.

Source:MENAFN.https://menafn.com/updates/pr/2020-01/28/AS_9bb67daf-cimage_story.jpg

In this section, an outline of algorithms specialized in autonomous navigation in an unknown environment is presented. Various procedures like mapping, localization, and path planning are involved in exploration. A map must be made to control the development of a robot in a specific situation. The versatile robot operating in an obscure domain and gathering helpful information to develop a map for further navigation is called autonomous exploration.

This chapter is structured in the following manner: Section 12.2 describes the algorithms used for autonomous navigation in the proposed work. Section 12.3 covers the design of the proposed hybrid model for an autonomous robotic system in a dynamic environment. Furthermore, Section 12.4 illustrates the past studies proposed and implemented in various application areas by different researchers. Moreover, the last Section 12.4, delivers a summary of the chapter.

12.2 ALGORITHMS FOR AUTONOMOUS NAVIGATION IN THE PROPOSED WORK

Conventionally, robots have been fully tele-operated by human administrators. However, with an increase in applications of the robots and the diverse scenarios in which they have to act, the self-learning route can expand the scope of the achievement. The algorithms intended for path planning consume more energy, decreasing the rate of the robot. Traditionally, obstacle avoidance and path planning were performed in 2D using 2D sensors. Since it is not precise to represent the 3D world in 2D images; hence 3D images can provide better mapping and navigation for the robots. The path planning problem is a well-known example of an NP-hard problem, and the complexity rises as the vehicle's degrees of freedom increase [3]. From the past studies, it appears that there were algorithms designed for autonomous navigation, such as Dijkstra, A*, D*, Genetic Algorithm (GA), Simultaneous Localization and Mapping (SLAM), Simultaneous Positioning and Mapping (SPAM), Reinforcement Learning, and the list continues. The impending segment of this chapter elaborates on reinforcement learning followed by cuckoo search optimization technique, and symbolic computing will be discussed in the latter part of the chapter.

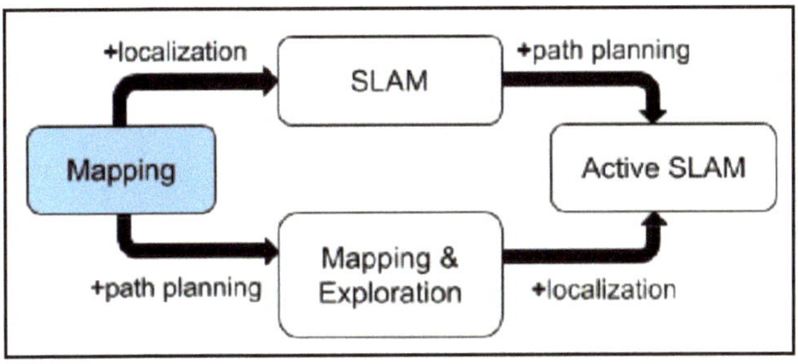

12.2.1 Q-LEARNING ALGORITHM

Reinforcement Learning is a goal-directed machine learning system. The task is started by interaction with the environment, followed by action. Every control issue that iteratively communicates with the state of the climate typically includes an RL control agent. The behavior of actions adapts to changes in signals and the length of the activities. This adaptation is determined by behavior policies π and feedback reinforcement rewards r, which are estimated based on wait time, delay time, transit time, subsequent.

States s,' and the transition probability P optimized by the RL agent. The reinforcement learning policies are created by linking possible states (S) and actions (A) using a learning factor, discount factor, and rewards. The policies are continually adjusted with the rewards received during the learning process until the desired outcome is achieved. Figure 12.4 portrays the illustration of the learning process in reinforcement learning.

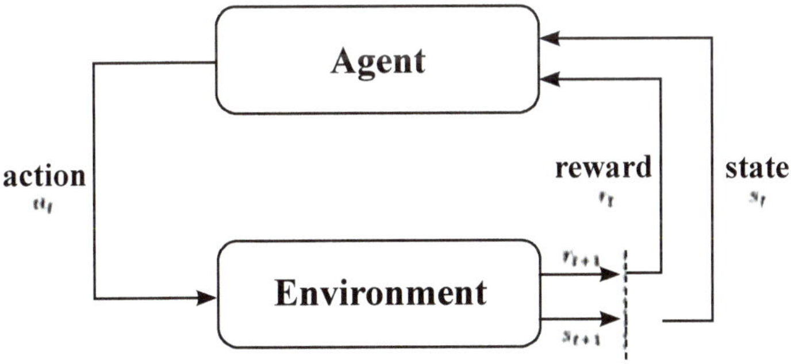

FIGURE 12.4 Illustration of the learning process.

The recognition of control is determined by a combination of state and action. In Reinforcement Learning (RL), the calculation of the value function is based on the rewards obtained from each state. The value of a state is determined by the accumulated discounted reward over a long period of time, based on policy π, for a given state 's.' It is given as:

$$V\pi\ (s) = E\ [\ Rt\ st = s] \tag{1}$$

Eqn. (1) is viewed in the bellman form, and it is given as:

$$V\pi\ (s) = \sum a\ \pi(a \vee s) \sum s,r\ P(s, r \vee s, a)[r + \gamma V\pi\ (s')] \tag{2}$$

The concept of an action's worth is referred to as a 'Q-value' which is comprised of the long-term discounted reward for the selected action 'a' and state 's' based on the policy π. This can be expressed as: $Q\pi\ (s, a) = E[Rt \vee st = s, at = a]$.

This value is further broken down into the Bellman form:

$$Q\pi\ (s, a) = \sum s',r\ P(s', r \vee s, a)$$

Q-Learning, which operates off-policy, is a well-known Reinforcement Learning algorithm. In Q-learning, the exploration of the learning process of actions is done independently of the policy. To choose an action 'a' from the possible actions A, the RL agent relies on the highest Q-value, which is determined through a process of greedy search. The Q-values are saved in a matrix called the 'Q-table,' which contains discrete values. This table computes the value of a state 's' from the set of states S, along with the value of an action 'a' from the set of actions A. The updated Q-values during the computation process are learned through the following equation:

$$Qk\ (s, a) = (1 - \alpha)Qk - 1\ (s, a) + \alpha$$

Here, Qk represents the updated Q-value at the kth learning step, Qk1 is the existing Q-value stored in the Q-table, r is the current reward value at the kth step of the learning process, s is the current state value at the kth step of the learning process, a is the current action value at the kth step of the learning process, and α is adjustable learning rate [1]. The detailed process of computing rewards, states, actions, and the learning rate will be discussed. Since the Q-learning algorithm performs both on-policy and off-policy, it is collectively grouped under the class of Markov Decision Problems (MDP). The MDPs are defined as making decisions in a sequential process based on the selected controlled actions. Stochastic control theory mostly resolves these problems. With the baseline of the work explored by Bellman, the computational complexity of the MDPs is adjudged with the help of Dynamic Programming (DP). It is primarily applicable to large-scale and complex problems that include discrete steps such as Policy Iteration (PI) and the Value Iteration (VI). These are manipulated, preserved, and estimated by the concept of Transition Probability Matrix (TPM). Since the random variables are instantiated, it is quite complex to deal with the probabilities of the transitions. This problem is known as the curse of modeling. The transition elements are more significant in dimension, which impacts the storage and manipulation process. This problem is stated as the curse of dimensionality. Both problems are challenging in the DPs, which is more reflected in the application areas of the robotic systems. The system under the MDPs is treated as the chain of action process. In the perspective of the Markov chain, the system navigates randomly from one state to another state at a discrete time. In continuation with the previous states, the current transition will be determined. The current state is determined by the newly updated learning value. The system is in charge of choosing the actions from the list of system actions for the subset of states [4].

A policy is established for every state. These states are rewarded with a reward value r, positive, negative, or zero. It helps the system to select the best actions. Therefore, the task of MDPs is to find out the optimal policies that give an accurate performance measure. Overall, it contributes more to the discrete time taken for the learning process. It has two well-known metrics: the average reward and the discounted reward. The discounted reward is the aggregate sum of the rewards earned over a given discrete period of time in pursuit of policy, whereas the average reward is the most anticipated reward gained at each step of the learning process.

Let d(i) represent the selected action in state 'i' for the policy 'd.' 'd' is comprised of a finite set of states, |S|. The expected reward from state 'i' to state j for action a is represented by r(i, a, j), and the probability of the same transition is represented by p(i, a, j).

- **Definition 1:** The policy d' under the discounted reward at the state 'i' is given as: $Jd'(i) \equiv \varepsilon$, where, xs is the state prevailed before the $s\ th$ transition; γ is the representation of the discount factor; ε is the expectation operator, which will be updated to derive the optimal solution [6].

- **Definition 2:** The average reward of the policy d' at state 'i' is given as: $\rho d'(i) \equiv \varepsilon$. The representation of symbols is above the same, in specific to, the average reward is unaffected by the initial state [7]. By deviating the learning rate, an optimization problem is experienced under the learning process. Henceforth, the findings of the underlying optimization problem become a vital part of the robotic systems.

12.2.2 CELLULAR AUTOMATA

The robotic systems are used for automation processes in various application fields. In manufacturing sectors, some robots are employed for navigation purposes, i.e., randomly moving from one location to another. In some cases, multiple robots are needed to speed up the navigation process. Different tasks need to be processed in such scenarios. When a robot is assigned a task, it moves randomly from its initial position to its destination within a defined working area. Therefore, the navigation control of robots is known as path planning, which is of prime importance. The goal of moving robots has become an interesting research topic among experts, focusing on planned paths. Constructing paths and navigating robots along these planned paths are challenging tasks. Cellular automata are used to build obstacle-free optimal paths between the source and the destination. During the process of following planned paths, controlling the torque movements becomes critical. In the path planning process, designing the positions of the robots can create obstacles in the working area. Based on the presence of the information, the planning of the paths may be static or dynamic. The information about the obstacles before the navigation control of the robots is known as static path planning. The partial information about the obstacles given to the navigation control of the robots is known as dynamic path planning. The off-line process refers to the static process, whereas the on-line process refers to the dynamic path planning. Several methods have been developed to identify obstacles at an earlier stage to avoid delays during navigation. Some authors have developed graph-based approaches that connect the source and the destination with the help of intermediates. The edges of the source and the destination points are connected with a possible set of obstacle-free paths. A path generally represents a straight line that interlinks the sub-points of the vertices among the source points, destination points, and obstacle points. It is found that the intermediates cause more obstacles before reaching the destination point. A visibility graph is constructed to identify the available paths, and from these, an optimal path, i.e., the shortest path, is discovered. The shortest path is the one that holds a minimized distance cost, which is the prime motive of the research study. Once suitable paths are discovered, the optimal path is selected by different searching algorithms. Along with the graph-based approaches, cell-based decomposition approaches are also established for an optimal path planning process. CA is a branch of science that uses digital processors to represent complex processes with discrete characteristics. It is a recent concept in the study of autonomous robot systems, which are mostly decentralized. Cellular Automaton (CA) generates new transition rules by considering how neighboring cells interact and what states a cell can potentially be in. By locally interacting with the components of navigation systems, it creates a dynamic system. The system starts with N cells arranged in a lattice, each with a unique identification pattern in its connections, which are used for communication with other cells. A transition rule is calculated for each cell state at time t+1 based on its interaction with neighboring cells. The finite sets of the d-dimensional lattices are represented by $Z\ d$ and the finite fields are represented by Σ. The communication between cells in the lattice is usually shown in a map form as c: $Z\ d \rightarrow \Sigma$ [8]. Mathematically, it can be expressed as $si\ t \rightarrow$ At time t, the state of cell 'i,' represented by $si\ t$, is part of $si\ t\ \varepsilon\ \Sigma$. The states of the current and neighboring cells are stored as $\eta i\ t \rightarrow$. The transition rule of a cell in the lattice is given as Φ: Σ n $\rightarrow \Sigma$ [9], where the size of the neighborhood is represented by n and $\Phi(\eta i)$ gives the information about the neighborhood state that affects cell 'i' at the next time step, $si\ t$+1, as a function of $\eta\ t$. These features of CA assist in the path-planning process by considering the attraction to the goal and local decision-making. The CA rules, including the transition rule, are used by the robot's sensor to compute the neighborhood's local decision and state at each discrete step. The aim of this work is to guide deployed robots to reach their goal point in a straight line, even if obstacles are encountered. To achieve this goal, a team of robots has been formed with a desirable navigation pattern. The robotic environment is a two-dimensional square grid with each cell having a side length of l, and the robots may be larger than the cell size, but they are still considered as one cell. The robots use infrared proximity sensors to gather information about the state of the surrounding cells. Two navigation patterns, triangular and linear, are used to assess this information. The robots' transition rules are controlled through two main rules: the first rule deals with deviating from obstacles, while the second rule manages the navigation. A master-slave control architecture is used to manage these rules, where the primary robot gathers information about the positions of other robots on its team and determines which rule to implement. Thus, this process requires a decentralized control system, which is achieved through CA-based approaches. The aim of the work is to guide robots to reach a target point in a straight line while avoiding obstacles along the way. A desirable navigation pattern has been established as a team effort to overcome the obstacles. The robotic environment is in two dimensions and consists of cells with a side length of l. The robots, even though larger than l, are considered as single cells. The state of the surrounding cells is detected using infrared proximity sensors. The navigation process is evaluated through two patterns: triangular and linear. The control architecture is based on two transition rules, the first one being how to deviate from obstacles and the second one being how to control the navigation. To make decisions about further processing, the primary robot gathers data on the locations of other robots on its team and chooses the appropriate rule to apply. To facilitate this decision-making process, a decentralized control system is used, which is implemented through the use of CA-based methods. Two strategies are followed in the path planning process: building of neighboring cells, decision-making process of transition rules, and applying transition rules on robot actions. The robots can exist in one of three possible states: Free (Fr), Robot (Ro), or Obstacle (Ob). The neighborhood cells are established based on the sensor readings obtained from the environment. The information about the neighborhood cells is preserved in vector form as, V= $\{v0, v1, v2, v3, v4, v5, v6, v7, v8\}$ and ΘR stores the current angle of rotation. It is claimed that the five rotational angles are possible. Consider an instance that a robot is willing to navigate south to the north. Then, the possible angle of rotations is computed as, and shown Figure 12.5.

v_8	v_1	v_2
v_7	v_0	v_3
v_6	v_5	v_4

FIGURE 12.5 Spatial neighborhood.
Note: The initial points (a) v_0 to v_1 is 0^0, (b) v_1 to v_2 is 45^0, (c) v_2 to v_3 is 90^0, (d) v_3 to v_8 is -45^0, (e) v_8 to v_7 is -90^0.

During a specific time, step (t), a robot can perform one of two types of discrete movements. The first movement involves navigating to the next cell by monitoring the orientation of the cells in its current position. The second movement involves rotating the current orientation of the cells. During the movement process, the distance is calculated as 1 for neighboring cells and √1 for diagonal cells. Each cell is updated by its current states, just by defining the next movement of the robots using the transition rule of CA. Overall, a transition rule is made up of the current state of the neighborhood and the angle of rotation, ΘR. With the help of a set of rules that are generated, such as the control rule and the deviation rule, each robot knows where it is and gets to the goal points. If something gets in the way, another deviation rule will be used. Else, the control rule will be used. 26 $v8$ $v1$ $v2$ $v7$ $v0$ $v3$ $v6$ $v5$ $v4$. The deviation rule is made up of two sets of rules. One rules how the next cell, 'i' moves, and the other rules how the robot moves. The robots follow a set of rules that tell them where to go. The first rule states whether or not the robot is free inside the cell, and the second rule says how the robot should move around if it is free. The CA rules are created to deploy multiple robots, and the formation control rule guides the robots to their target points each time step progresses. If an obstacle is encountered, the formation control rule can help the robot navigate around it by reversing its orientation or returning to a previous path. If the robot is on-axis, it moves forward, but if not, it takes diagonal steps. However, the challenge arises when applying the deviation rules when an obstacle is detected in the intended path [1].

12.2.3 CUCKOO SEARCH OPTIMIZATION

The cuckoo search optimization is responsible for selecting the optimal paths from the available paths learned from the Q-Learning and cellular automata processes. Optimization is crucial for efficient autonomous robots, as the environment is unknown and contains a large number of obstacles. Selecting the optimal path in an unknown environment is a difficult task. To ensure that the robot reaches its goal without any obstacles, a different sensor is deployed in real-time environments [1]. It is a time-consuming as well as a cost-efficient process. The planning of the paths may be local path planning (or) global path planning dependent on the surroundings. The building of global path planning demands a completely known environment, whereas the local path planning performs even in known and unknown environments. Through various approaches suggested for efficiently finding the paths, it is in under a developmental stage. The shortest path is the possible path free from obstacles and constraints of the environment, and a smooth straight line is required between the initial point and the goal point. This kind of scenario is projected as an optimization problem. Here, a recent optimization algorithm known as cuckoo search is explored for the path planning process. The Cuckoo Search is a type of metaheuristic search model. Researchers developed it based on the swarming behavior of birds, specifically the way cuckoo birds lay their eggs in the nests of other birds. This algorithm is simpler compared to the Genetic Algorithm (GA) and Particle Swarm Optimization (PSO) as it requires fewer simulation parameters. Cuckoo birds follow an aggressive reproductive method, laying eggs in other birds' nests and sometimes causing destruction to the nest or eggs. This optimization model uses the reproductive behavior of cuckoo birds and their Levy flight behavior, which is a type of random walk process, to generate new nests at each iteration. The probability distribution of the levy flight follows a power law, represented as $y = x - \beta$, where; β is a variance that ranges from 1 to 3 (Figure 12.6). The rules for deploying Cuckoo Search are:

- At any given time 't,' a cuckoo produces a single egg and simultaneously places it randomly into another nest.
- The nest that contains the highest quality eggs is considered the best and is passed on to the subsequent step.
- A fixed number of host nests are assumed, and then the egg found in the nest of the host bird is calculated as a probability $pa\varepsilon$ [1].

In some cases, the host birds might build new nests in different places. If this happens, rule C could be maximized or minimized based on random solutions. In a maximization problem, the size of the fitness value is related to the size of the goal value. In other words, the solution represents an egg in the nest, while the new solution represents the cuckoo's egg. The objective of this system is to achieve a better solution for a nest. This can also be expanded when dealing with multiple eggs in the nest. A new solution generated by levy flight behavior is given as: $Xi\ (t+1) = Xi\ (t) + \alpha \oplus levy(\lambda)$. The variable α adjusts its size depending on the defined problem area, and is often assumed to be 1. This equation is based on the Markov chain process, which describes a random walk, meaning that its

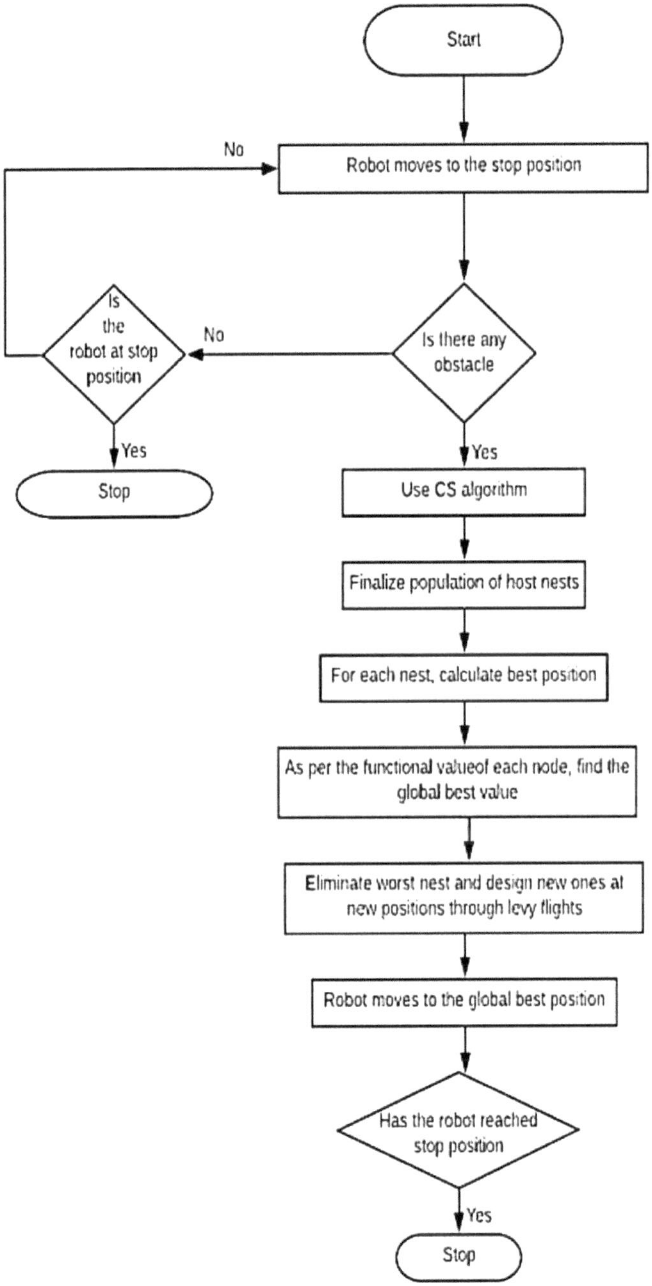

FIGURE 12.6 Flow chart of cuckoo swarm optimization.

previous location determines the following location of the egg ($Xi\,(t)$) and its probability of transition ($levy(\lambda)$). The $\oplus$ operator is an entry-wise multiplication that helps to explore the search space for a longer time [1].

12.3 DESIGN OF THE SYSTEM

The design of the suggested hybrid model for an autonomous robotic system in a dynamic environment is covered in this section. The assumptions needed for the system design are covered.

12.3.1 EXPLORATION OF TERRAIN AND THE COVERAGE OF THE ROBOTIC ENVIRONMENT

Currently, the majority of robotic applications require that robots intelligently choose the best course by investigating and absorbing their surroundings. At the same time, by removing the impediments, the robots should be able to arrive at the goal spots at a specific discrete time. In some circumstances, they ought to be able to respond to dynamic requests with the aid of prior knowledge and experiences. Multi-robot deployments in an unknowable environment that give a higher level of scalability, resilience, and a reduction in redundancy are made possible by implementing technical advancements. In order to create a collaborative atmosphere, path exploration and coverage are also properly managed. The robot may learn about its surroundings depending on the area covered by the installed sensors, and it is also not given any prior knowledge. Robots and humans both engage in various types of exploration while detecting their surroundings. Despite that, exploration covers the borders of the environment, even at remote sensing, whereas coverage senses the working area of the environment. Instead of optimizing the amount of actions in a sequence, it was necessary to build the proposed algorithms to suit the needs of multi-robot performance enhancement. An effective path planning procedure can be built with the aid of coverage analysis. The cellular breakdown of the environment should be used in conjunction with the identification of the geometrical characteristics of the coverage region. After the cells have been divided up, search-based coverage techniques are implemented. Due to the arrangement of the furniture and the fluctuating human behaviors in an unstructured (or unknown) setting, the coverage algorithm becomes very difficult [23].

12.3.2 MODELS AND THE ASSUMPTIONS

The robots only understand their starting position. They must have sensors installed that can automatically identify obstacles in their environment. The unknown environment was divided into grid cells with a size of 20 cells. The robot's internal sensors picked up on these cells. The cells of the whole area were covered when the robots were placed in the center of the region. Additionally, it was believed that the sensor range of the robot would be able to detect obstructions in the three-dimensional orthogonal perspective of the surrounding cells. Each robot in a cell underwent a unique spanning tree administration during the course of the robots' cell exploration. The algorithm used the cell's current location as input, and the result was the robot's spanning tree. Since it scans using orthogonal perspectives, every position—including the top, bottom, up, and diagonal—was explored. The positions of the robots (RP), obstacles (O), and intermediate cells are shown in Figure 12.7, which shows the sensing orientations of the cells (a, b, and c).

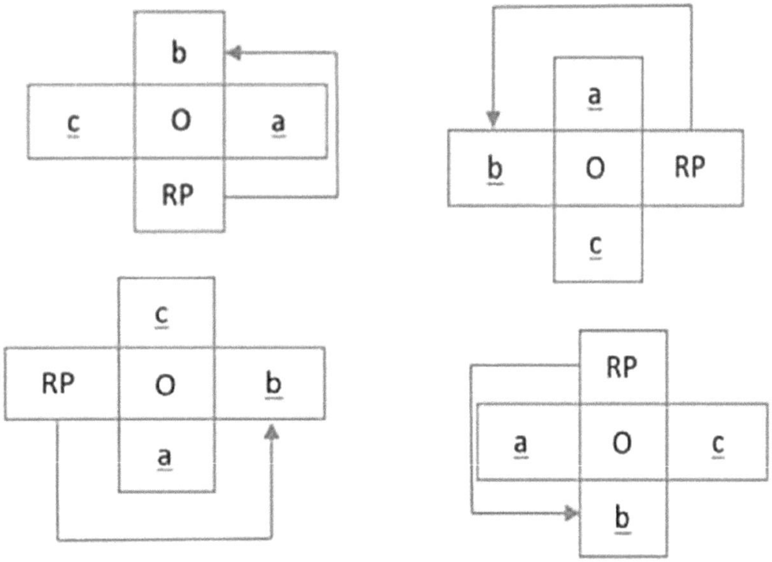

FIGURE 12.7 Sensing direction of the cells.

Here, the robots can name the cells they have visited, allowing for the separation of the robot cells that are covered from those that are exposed. The robots also covered the sub-cells in a predetermined terrain with the aid of intermediate cells. The Q-covered table's cells were constantly updated. By consistently pushing the cells into the stack S, this table makes it possible to quickly cover the exposed cells. Our findings show that later, the covered cells were taken up for the additional analytical step via path planning. The cells could be accessed from any point because the deployed robots were uniform in character. It was also anticipated that since decentralized algorithms were created, robots would implement them freely.

The suggested approach had limitations for an unknowable environment. First, a continuous, planar terrain structure with an exterior boundary was taken into consideration. Moreover, it was divided into 2D square cells with four sub-cells. Moreover, robots were continuously navigating between two nearby sub-cells. Furthermore, it follows a full path-coverage, meaning that the starting and ending points of the sub-cells stay distinct. Last but not least, Robot only makes one motion at a time.

12.3.3 A HYBRID MODEL FOR AUTONOMOUS NAVIGATION

The suggested system uses the three terminologies we previously covered for autonomous navigation and path planning: a "new optimization employing Q-learning" that aims to create a robot's ability to choose the best shortest path in a dynamic environment. The suggested plan functioned as a sort of self-learning system. The method operates progressively by creating several Q-tables for numerous robots at the same time. Each robot was presumptively started from its starting points on the terrain (S1, ..., Sk). The goal of this study was to determine the best shortest path starting with the k-robots' k-sub-paths. On the other side, it was carefully observed that the robots had covered the entire landscape. In a similar manner, the robot only passes over each of its sub-cells once. The robots' navigation method involved some controlled activities that are referred to as behavioral features. Here, four behaviors were identified, including cell selection, border recognition, obstacle avoidance, and mobility. The suggested plans are described as in subsections.

12.3.3.1 Q-LEARNING ALGORITHM

This algorithm aims to learn about the behavior of the cells and their subcells during the navigation process of robots. To make an efficient sequence of decisions, this algorithm takes trial and error on the defined parameters.

The deployed robots initiate the working process by just learning the environment via reward and penalty process. Here, the selected action was accurate, then the learned data was represented as "reward" else it was labeled as "penalty." This learning process was done by the Markov Process and thus, decisions made from this process are known as "The Markov Decision Process." The term „Markov" was coined because it portrays the future states based on the information provided by the present states. Let S = {S1, …, Sk} be the set of states and t be the present time of the states. Then, the probability state of Markov process was given as, P [St+1|St] = P[St + 1|S1, S2, · · ·, St].

Since the robots were homogeneous by nature, the Markov chain under the transition probability was independent towards time t, which was given as P[St+1 = s'|St = s] = P[St = s'|St–1 = s]. Let the tuple (S, P) be the Markov chain process, in which S represents the finite set of states map and P is the transition probability matrix of the state. The objective of the proposed Q-learning algorithm was to maximize the rewards of a robot by taking a consecutive series of actions in the dynamic robotic environment. The rewards were maximized by learning the environment and representing it in the form of transition maps [22] (Figure 12.8):

- **Step 1:** Initializing the Q-Table: The learning process was started by creating the Q-table. It was projected as an m * n matrix, where m represented the number of states and n represented the number of actions. It consists of four actions: via, up, down, left, and right, and five states: via, start, idle, correct path, wrong path, and the end.
- **Step 2:** Selecting an Action: Depending on the robotic environment, the learning process begins by selecting any one action at time t.
- **Step 3:** Performing an Action: Here, the actions were performed based on time steps. The robots learn the environment by collecting the previous and past states of the cells by taking a series of actions. Initially, the Q-value in the Q-table was represented to be 0. Consider an instance, (action a) right () performed to start (state 's') the process, then the Q-table was updated using the Bellman equation. The exploration process begins by performing a greedy strategy on the dynamic environment. Therefore, the sample updated Q-table according to the Bellman equation was shown in Figure 12.8.

Action	Up	Down	Left	Right
Start	0	0	0	1
Idle	0	0	0	0
Correct Path	0	0	0	0
Wrong Path	0	0	0	0
End	0	0	1	0

FIGURE 12.8 Updated Q-table.

- **Step 4:** Measuring the Rewards: This was the crucial step that helped to build an efficient path planning. It was evaluated by the function Q(s,a). The learning rate was determined at the initial step. The function Q(s,a) will continuously learn the environment until the criteria of learning rate terminates. The maximized Q-value of an action at that state was rewarded. The Bellman equation was given as:

$$Q^* (s,a) = Q\,(s,a) + a[R\,(s,a) + \gamma maxQ'\,(s,a) - \ (s,a) - Q\,(s,a)]$$

where; Q* (s,a) → Updated (new) Q-value for the state and action; → Present Q-values; → Rate of learning; → Rewards taken for an action at a state; → Rate of discount; → Maximum Expected future rewards; Here, the rewards are symbolized as follows, (+1) → When the current step was closer to the goal; (–1) → When the current step hits the obstacles; (0) → When the current step was in idle state.

Figure 12.9 represents the sample final Q-table. In this way, the Q-learning algorithm has helped to find (or) learn about the unknown environment. In the study, it was known for collecting the data for the training process using the Q-learning algorithm. The estimation of reward value portrays the robust learning model that enhances the performance of the system. Likewise, the discount factor helps to estimate the data by an agent which was greatly helpful to deal with the shortest path planning. It reduces the effects of data overfitting (Figures 12.9 and 12.10). Figure 12.10 presents the workflow of the Q-learning algorithm.

Action	Up	Down	Left	Right
Start	0	0	0	1
Idle	0	0	0	0
Correct Path	0	43	12	0
Wrong Path	12	0	23	0
End	0	0	1	0

FIGURE 12.9 Final Q-table.

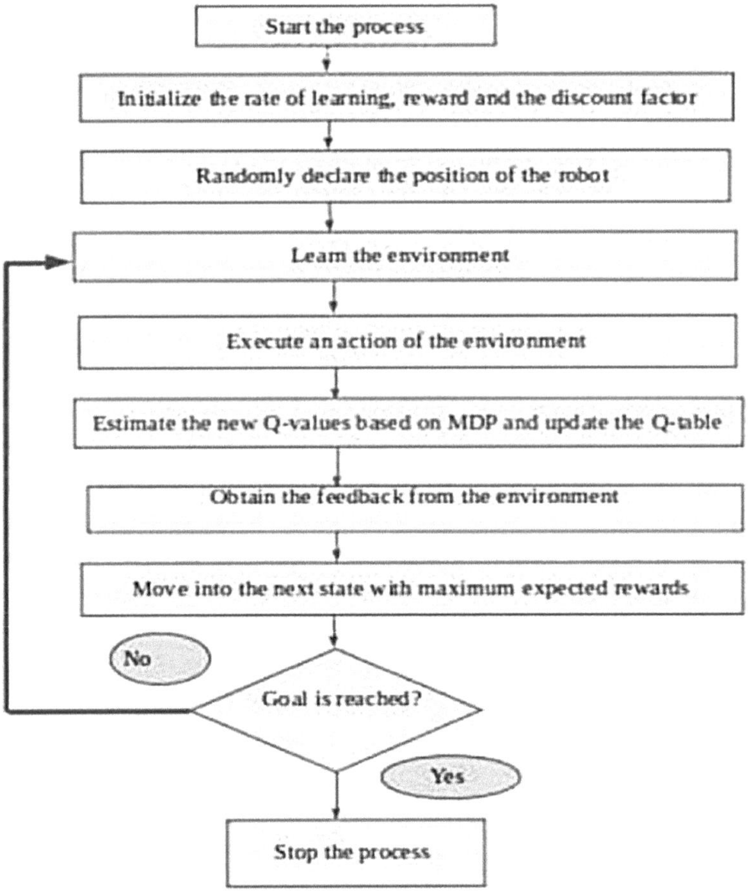

FIGURE 12.10 Flowchart of the Q-learning algorithm.

12.3.3.2 CELLULAR AUTOMATA AND CUCKOO SEARCH OPTIMIZATION

Once the training data process was created, the position of the robots was explored by the concepts of cellular automata. Here, the objective of the study becomes more complex because of finding the optimal actions from the set of actions. Since the environment was projected into a grid of cells, each cell contributes as a learning component. During the learning process, the best actions were learned to recommend the neighboring components for further actions. In our study, an optimal action selection from a set of actions was done by a parallel reinforcement learning process. By doing so, the system was able to manage all kinds of actions as well as preserve the time and space complexity of the environment. Once the optimal action was selected, the optimal shortest path was analyzed using CSO. Initially, the position of the robots by performing an action in a state was explored using cellular automata. It was a hybrid learning model that combines the learning process and the automation process of a cell of a robot in a dynamic environment. With the help of Q-learning, the automation process by determining the local rules. The objective of this study was to explore the fullest capability of maximizing the expected reward function for each learning automation process. Let the aggregate number of actions at cell 'i' be and the action to be chosen to be. Then, the set of actions of cell "I" was denoted as. Let be the policy strategies associated with it and be the number of neighbors of cell 'i' [24].

At step k, the probability of selecting was given as and the policy strategy over, and the average actions of reinforcement learning at time step k was given as. Let be the index value of the neighboring cells. The learning system operates on the probability of the set of actions. Initially, the joint distribution of the actions was estimated by the maximum empirical rewards. It conjoins with other learning automata even on neighborhood cells. The obtained CA combines with the probability of joint distribution, so as to form a single optimal action (or) superaction. The action increases gradually by their selected probability. During the exploration process, the probability of the superactions was acquired. Each selected action in the neighborhood was continuously observed and their empirical values were updated frequently. Every performed action sent feedback from the environment to its learning component. Depending upon the obtained feedback and the probability distributions, the estimated

Q-table was updated. Based on the selection, a mutated greedy selection process was done. Initially, the maximum expected reward action must be selected. Here, let us assume be the present action. Then, the new action from the selected action was computed as:

$$p_l = \frac{1 - q_l(k)}{\Sigma_{j\in I} 1 - q_l(k)} \quad \forall l \in L$$

Where,

$$q_h = \frac{Er\left(a_{ih'} \underline{pn}_i(k)\right)}{\Sigma a_{is\notin ai(k)}\, Er\left(a_{is'} \underline{pn}_i(k)\right)} \forall h \in \{l / a_{il} \in A_i \wedge a_{il} \notin a_i(k)\}$$

The above equation was continuously operated for the maximum expected reward of superaction. Then, the reinforcement learning vector:

$\beta_i(k)$ was calculated for the chosen superaction $= 1$, as below.,

$$p_{iJ}^{\wedge} + 1) = \left\{(1 - \gamma(k))\, p_{iJ}^{\wedge}(k) + \gamma(k) if\ a_i(k) = a_{ij}\right\}$$
$$\left\{(1 - \gamma(k))\, p_{iJ}^{\wedge}(k) otherwise\right\}$$

$$Q_i(k+1, :) = Q_i)$$

$$Q_i\left(k+1, a_{ij}, a_{(\,)}(k)\right)$$

$$Q_i\left(k + a_{ij}, a_{(1)}(k) .... a_{(\,)}(k)\right) +$$

$$\eta(k)\left(\beta_{ij}(k) - Q_i\left(k, 1, a_{ij}, a_{(1)}(k)., ... a_{(\,)}(k)\right)\right)$$

$$\forall j \in \{l / a_{il} \in a_i(k)\}$$

The above process was continued until the stopping criterion was met. During each time step k, the selected superaction was learned *via* learning component i.e., neighborhood cells. Since each action was performed separately, the learning process became more viable [22, 23].

The probability of the empirical distribution of all super actions was then given as input to the optimizing system. The constructed maps assisted in tracking their position and planning the navigation time. In order to achieve the main goal of this study, this optimization technique has been very helpful in our unknown environment. The mathematical procedure discussed above leads to the self-learning process for autonomous navigation of the mobile device. During the implementation of the above process, the following things were inferred. Firstly, the robot can attain its target point, even in the presence of obstacles. In some cases, dynamic obstacles are also created. Thus, the formation of a rectangular shape of the grid points has helped the robot to find its paths, so as to attain the targets. Also, the deployed robots can also eliminate (or resolve) the obstacles from the opposite directions, i.e., obstacles presented in the inner boundary as well as the outer boundary. Further, it ensures that the robot takes obstacle-free movement. It also ensures that the shortest path is being taken and thereby, the minimum cost and minimum time are also guaranteed. In addition, it is really helpful for all sorts of dynamic environments. The proposed scheme takes discrete control over the robotic learning agents, which is helpful for an efficient dynamic path planning process. To begin this, path planning is carried out on some barrier constraints known as obstacles. With the help of discrete time steps, the robots are permitted to navigate on the planned paths. Likewise, the learning rate is defined to be less than the sensing range of the sensor associated with the robot. At each step, robots look for the availability of obstacles in the navigation process. Along with that, the position of the obstacles is also estimated. If the robot finds any obstacles within its distance step, then a new obstacle-free path is suggested by the cuckoo search model. Algorithm: Input: A= No. of actions, S < R: No. of actions to be selected, $(\lambda(j))\_j\epsilon N$: Rate of the learning, $(\eta(j))\_j\epsilon N$: Rate of the averaging factor, $(\gamma(j))\_j\epsilon N$: Rate of the discount factor.

Output:

1. While (Maximum Generation < Stopping criteria), do
2. For each cell C, do
3. Explore the cells with $(\lambda(j))\_j\epsilon N$
4. Select the best action as α_i (k)
5. Performing greedy selection process to obtain maximum expected reward for the action α_i (k)
6. Select the reinforcement vector for all α_i (k), Set $k \leftarrow k+1$
7. End for
8. Do, Finding the set of paths by observing the cells.
9. For Robot Cell (C) $\neq$Target Cell (T)
10. Find the best cell
11. Evaluate with the free cell
12. End
13. Finding the paths, Do
14. Initiating the set of paths as Cuckoos
15. Defining objective function, f(N), N = (N1, N2, ..., Nd)T

16. For (F_i == Best (f(N)), F_i ++)
17. Obtain the best solutions, i.e., best path (P)
18. End for
19. End do.

Here, the point of the obstacle and the present position of the robot were declared as the starting point. Now, again the path was planned between the new starting point and the target point in connection with new intermediate points. Then, an optimal distance value was estimated, only if the robot does not face any obstacles during the navigation process. Here, the path exploration process was done by two processes, namely, local navigation and global navigation process.

12.3.3.2.1 Local Search Navigation Process

Here, the robot was assisted with a straight line that contains the starting point and the target point. This path was followed by the robots until it recognizes the obstacles. The detected obstacles can be resolved by deviating from the current position of the robot, which was frequently updated by the learning component. Then, a new position and new distance will be updated for assisting the upcoming robots. In this process, the position of the target point was already known to the robots.

12.3.3.2.2 Global Search Navigation Process

It was a type of planning process where the robot moves randomly with the prior information known to them. It was the process of choosing the best way to reach its target points by information loaded in it. Thus, for global path planning, the decision of the moving robot from a starting point to a goal was already made and then the robot was released into the specified environment.

12.4 THE WORK FOCUS ON AUTONOMOUS NAVIGATION IN THE PAST

The literature on autonomous navigation and path planning algorithms illustrates that it is still a domain of research. The researchers are still focusing on the development of an efficient and effective algorithm for navigation. These procedures are regularly required for the arrangement of the local minima issue. Researchers are analyzing distinct, efficient ways in which this problem can be resolved. So, the recent works on robot navigation are discussed in relation to various domains in the following section [23].

One of the papers presented an optimized algorithm for navigation of a mobile robot in a 3D environment, which involves the integration of multiple methods to ensure efficient path planning and accurate spatial representation. The algorithm uses Dijkstra's algorithm to find the shortest path, and also considers the size of the robot relative to the size of obstacles in the environment. The challenges in dynamic motion planning for mobile robots were tackled by utilizing a distinct technique that considers the behavioral dynamics from a control standpoint. The robot engages in a dynamic interaction with its environment, which includes the mobile robot's motion planning. This interaction dynamically models and controls the motion planning process. The behavior dynamics effectively address the issue of motion planning for mobile robots by transforming it into a manageable problem through optimization in the robot's acceleration space and incorporating an integrated planning and control framework. A novel approach to address the challenges of motion planning for mobile robots in dynamic environments was developed. The cuckoo optimization algorithm was used to optimize the path. A 31-feature vector was employed to reduce computational complexity, and another technique was used to minimize the dimension of the feature vector to further reduce complexity. On the other hand, a study introduced an efficient shortest path planning algorithm for planar mobile robots with a time complexity of O(4 x n) where n represents the geometric complexity of the non-dynamic environment. The algorithm employs a combination of limitations and the greedy method using the Dijkstra algorithm, resulting in a fast and easily improved shortest path. A path planning model based on the Bézier curve was also proposed. The control points in the curve impact both the path length and smoothness. A new specific Chaotic Particle Swarm Optimization (CPSO) algorithm was proposed to improve the control points of the Bézier curve, with two variations named CPSO-I and CPSO-II. The selected points create an optimal smooth path while limiting the distance between the start and end points. The conventional PSO algorithm was compared with the results from CPSO-I and CPSO-II to evaluate the CPSO calculation. Another robot path planning approach by combining a hybrid optimization algorithm (BPSO) with an accurate Voronoi boundary network was introduced. The proposed method includes a faster convergence speed and a lower failure rate compared to other algorithms. Two efficient, intelligent optimization controllers were also proposed, the Bat Algorithm (BA) and the Flower Pollination Algorithm (FPA), to be used by an autonomous mobile robot to find the optimal path in unfamiliar conditions. The FPA was designed based on the pollination process of flowering plants, while the BA solves optimization problems through echolocation and frequency tuning. A fitness function is used to evaluate the path-planning task of the mobile robot, taking into account the distance between the robot and obstacles or the goal, and ensuring obstacle avoidance. The mobile robot utilizes the values generated by the objective function to reach its goal while avoiding obstacles in unfamiliar environments. On the other hand, The Improved Particle Swarm Optimization (IPSO) was hybridized with the Improved Gravitational Search Algorithm (IGSA) for multi-robot path planning in cluttered environments. The IPSO-IGSA hybridization balances the search and overuse by incorporating social characteristics from IPSO into the IGSA movement and employing co-evolutionary techniques to improve the speed of IGSA and the particle positions combined with IPSO velocity. This algorithm reduces the maximum path length and arrival time of each robot at its destination in the environment. The robots make individualistic decisions by using the hybrid IPSO-IGSA algorithm to understand and communicate their positions in the world map. A modified ant colony algorithm for the path planning of a mobile robot in a perceived stable environment [22]. The modified ant colony algorithm expands the search range and reduces the impact of local minima, allowing for a quicker focus. The turning factor is also taken into consideration in the path optimization process. A new way to solve optimization problems using several evolutionary algorithms, such as Genetic Algorithms, Differential Evolution Algorithm, Gravitational Search Algorithm, PSO, and Bee Colony Optimization, for use in multi-robot path planning. The fitness function of the Genetic Search Algorithm consists of two parts to avoid the impact of collisions with static obstacles, including a function for choosing the next position on an optimal trajectory based on the estimation of speed, and another for restricting acceleration. PSO is designed based on Newton's law of gravity and motion. Different variants of the Gravitational Search Algorithm have been developed, and the algorithm has found applications in various fields. The aim of the multi-robot system is for the robots to reach their specific goals. To achieve this, a fitness function is created in the Particle Swarm Optimization (PSO) algorithm that decides the next position of the robots as they move towards their objectives. The path-planning problem arises when an iterative algorithm must determine the next position of all the robots while satisfying all the constraints in the multi-target task [4]. The algorithm repeats until each robot reaches its destination. In the PSO algorithm, new features have been added to improve its ability to determine the path direction for multiple robots, which uniquely defines starting positions to specific target locations in the environment to minimize the total path length of all the robots. The results show that the algorithm has improved the quality of path arrangement within a reasonable time frame. The IPSO algorithm is enhanced to handle the path-planning problem of the multi-robots by increasing the integration rate. Finally, the simulation has demonstrated the efficiency of IPSO with the Khepera environment, and the results are compared to other algorithms, including PSO and DE. The IPSO method is based on a dynamic structure and has strong performance, self-deterministic interaction, and works in a challenging domain in a multi-robot system. IPSO suggests a path-planning strategy that robots can use to get to their desired locations from where they are now in the environment. Two scholars came up with a new way to make sure a robot can move from where it starts to where it needs to go without hitting any obstacles. The technique makes the existing A* algorithm better, so the robot can move around in a place it doesn't know where there are fixed obstacles. The Enhanced A* algorithm provides an optimal path for the robot to reach its target. A systematic review of different artificial intelligence techniques used in the navigation of robots was conducted, including Neural Network, Particle Swarm Optimization (PSO), Fuzzy Logic, Genetic Algorithm, and other AI techniques. The review analyzed the role of various AI techniques in controlling and navigating different types of robots under various environmental conditions. A new approach by combining η3-splines with a modified particle swarm optimization (MPSO) was presented. To begin with, arbitrary points are connected using η3-splines, with kinematic parameters chosen to relate to the control and motion of mobile robots. The MPSO algorithm uses adaptive random fluctuations (ARFs) to control frequent local convergence and ensure smooth global path planning. The main challenge lies in arranging the path in a smooth manner. The MPSO algorithm incorporates an evolutionary factor, which evaluates the classification at each iteration, resulting in an evolutionary state. The velocity improvement dynamics are switched based on the evolutionary state with ARFs imposed on the global/local best particles based on the current iteration. Lattice-based modular robot networks with only neighbor-to-neighbor interactions. These networks develop into sparse and wide breadth network. Strict limitations are established for the diameter and breadth of these networks, addressing the critical design issue in complicated dispersed networks caused by the big diameter and enormous average distance in massive-scale lattice-based networks. New research looked at the different ways mobile robots can be used in WSNs. This research gives researchers an idea of how each category works, how different solutions relate to each other, accurate information, and an in-depth analysis. The different approaches are put into different categories based on how they are made scientifically, how they are used, and so on. Through an optimal robot path planning system, a map was made which planned the best routes for mobile robots. The system creates a grid-based map by combining data from different sources and fixed obstacles. Using a simplified neural network model, the best path is determined, and the mobile robot is constructed. The mobile robot scans its surroundings to detect any moving obstacles and avoids them. A collision avoidance algorithm was developed for mobile robots by integrating the characteristics of enhanced ACO and APF, and by developing a new GRU-RNN system model for dynamic path planning in a novel domain. The GRU-RNN system model comprehends the primary framework's arrangement method, leading to a general enhancement in performance due to the utilization of the improved ACO and APF algorithms [5].

In summary, this chapter explores previous research in autonomous navigation, including the various algorithms used and their advantages and limitations. The next chapter will focus on the design and development of the autonomous navigation model.

KEYWORDS

- **autonomous mobile robot**
- **autonomous navigation model**
- **autonomous robots**
- **global path planning**
- **neural network model**
- **optimal robot path planning system**
- **programmable robot**

REFERENCES

1. Niharika, S., Manish, P., Piyush, C., & Tanupriya, C., (2020). An optimized self learning algorithm for autonomous navigation in an unknown environment. *Journal of Green Engineering -JGE, 10*(7).
2. Agarwal, T., (2012). *Autonomous Robots and Their Types with Applications*. www.elprocus.com (accessed on 27 April 2024).
3. Marin-Plaza, P., Ahmed, H., David, M., & Arturo De La, E., (2018). Global and local path planning study in a ROS-based research platform for autonomous vehicles. *Journal of Advanced Transportation.*
4. Das, P. K., Sahoo, B. M., Behera, H. S., & Vashisht, S., (2016). An improved particle swarm optimization for multi-robot path planning. *International Conference on Innovation and Challenges in Cyber Security (ICICCSINBUSH)*, 20.
5. Zafar, M. N., & Mohanta, J. C. (2018). Methodology for path planning and optimization of mobile robots: A review. *Procedia Computer Science, 133*, 141–152.
6. Richard, S. S., & Andrew, G. B., (2018). *Reinforcement Learning: An Introduction* (2nd edn.).
7. Panda, R. K., & Choudhury, B. B., (2015). An effective path planning of mobile robot using genetic algorithm. In: *2015 IEEE International Conference on Computational Intelligence & Communication Technology* (pp. 287–291). IEEE.
8. Hosseininejad, S., & Dadkhah, C., (2019). Mobile robot path planning in dynamic environment based on cuckoo optimization algorithm. *International Journal of Advanced Robotic Systems, 16*(2), 1729881419839575.
9. Sun, L., Liu, X., & Leng, M., (2006). An effective algorithm of shortest path planning in a static environment. In: *International Conference on Programming Languages for Manufacturing* (pp. 257–262). Springer, Boston, MA.
10. Tharwat, A., Elhoseny, M., Hassanien, A. E., Gabel, T., & Kumar, A., (2018). Intelligent Bézier curve-based path planning model using chaotic particle swarm optimization algorithm. *Cluster Computing*, 1–22.
11. Mo, H., & Xu, L., (2015). Research of biogeography particle swarm optimization for robot path planning. *Neurocomputing, 148*, 91–99.
12. Ghosh, S., Panigrahi, P. K., & Parhi, D. R., (2017). Analysis of FPA and BA meta-heuristic controllers for optimal path planning of mobile robot in cluttered environment. *IET Science, Measurement & Technology, 11*(7), 817–828.
13. Das, P. K., Behera, H. S., & Panigrahi, B. K., (2016). A hybridization of an improved particle swarm optimization and gravitational search algorithm for multi-robot path planning. *Swarm and Evolutionary Computation.*
14. Zhang, Y., Chen, C., & Liu, Q., (2016). Mobile robot path planning using ant colony algorithm. *International Journal of Control and Automation, 9*(9), 19–28.
15. Sudhakara, P., & Ganapathy, V., (2016). Trajectory planning of a mobile robot using enhanced A-star algorithm. *Indian Journal of Science and Technology, 9*(41), 1–10.
16. Parhi, D. R., (2018). Advancement in navigational path planning of robots using various artificial and computing techniques. *Int. Rob. Auto. J., 4*(2), 133–136.

17. Song, B., Wang, Z., Zou, L., Xu, L., & Alsaadi, F. E., (2019). A new approach to smooth global path planning of mobile robots with kinematic constraints. *International Journal of Machine Learning and Cybernetics, 10*(1), 107–119.
18. Naz, A., Piranda, B., Tucci, T., Goldstein, S. C., & Bourgeois, J., (2018). Network characterization of lattice-based modular robots with neighbor-to-neighbor communications. In: *Distributed Autonomous Robotic Systems* (pp. 415–429). Springer, Cham.
19. Huang, H., Savkin, A. V., Ding, M., & Huang, C., (2019). Mobile robots in wireless sensor networks: A survey on tasks. *Computer Networks, 148*, 1–19.
20. Chen, Y. W., & Chiu, W. Y., (2015). Optimal robot path planning system by using a neural network-based approach. In: *2015 International Automatic Control Conference (CACS)* (pp. 85–90). IEEE.
21. Yuan, J., Wang, H., Lin, C., Liu, D., & Yu, D., (2019). A novel GRU-RNN network model for dynamic path planning of mobile robot. *IEEE Access, 7*, 15140–15151.
22. Katal, A., & Singh, N., (2022). Artificial neural network: models, applications, and challenges. In: Tomar, R., Hina, M. D., Zitouni, R., & Ramdane-Cherif, A., (eds.), *Innovative Trends in Computational Intelligence.* EAI/Springer Innovations in Communication and Computing. Springer, Cham. https://doi.org/10.1007/978-3-030-78284-9_11.
23. Singh, N., Prateek, M., & Chauhan, P., (2019). An efficient path planning algorithm for networked robots using modified optimization algorithm. In: *International Journal of Innovative Technology and Exploring Engineering* (Vol. 8, No. 12, pp. 2738–2743). Blue Eyes Intelligence Engineering and Sciences Engineering and Sciences Publication – BEIESP.
24. Singh, N., Prateek, M., Chauhan, P., & Choudhury, T., (2020). An optimized self-learning algorithm for autonomous navigation in an unknown environment. *Journal of Green Engineering, 10,* 3315–3334.

CHAPTER 13

Environmental Learning for Autonomous Navigation

SANJEEV KUMAR[1] and S. B. GOYAL[2]

[1]*School of Computer Science, University of Petroleum and Energy Studies, Dehradun, Uttarakhand, India*

[2]*Faculty of Information Technology, City University, Menara City U, Malaysia*

ABSTRACT

The first instruction of the chapter is coupled with the most recent findings as to environmental system learning functioning as a guided autopilot mechanism as this is followed by the issues with theoretical background and the implementation of the practical techniques, as well as the roundup of the most advanced modern methods of the emerging out of this. Environmental robots are trained to differentiate with their surroundings at first and understand as time goes by which can have them be aware of any unlikely circumstances and can of course as well, boost their performance through time. It is one of the most vital functions which allows autonomous control leading to the safe work and better outcomes irrespective of the conditions. First, it discusses several environment modeling algorithms, in which the leading technique for robots to be able to GPS the surrounding and serving to pinpoint the initial position of the robot for a map is Simultaneous Localization and Mapping (SLAM). Beyond these principal algorithms, which is natural, include other algorithms as neural network methods, reinforcement leaning and Bayesian method. At this chapter, we will have an analysis of

Advances in Autonomous Navigation Through Intelligent Technologies.
Niharika Singh, Thipendra P. Singh, Brian Azzopardi (Eds.)

multiple mixed methods that are a result of combining different techniques to overcome each of the individual weakness that they suffered from. Data are sought by designing new data structures, evaluation of multi-modal fusion sensor, and generative models as the advanced learning algorithms in environmental systems in the contemporary times. This part ends with two lists: the author selves can implement simulated scenarios with the robots taking part in various datasets for the research on the simultaneous localization and mapping (SLAM).

13.1 INTRODUCTION

Applications ranging from self-driving cars and drones to mobile robots used in industry and healthcare highlight the importance of autonomous navigation in robotics and other fields. Autonomous systems need to be able to sense and interpret their surroundings, plot out efficient and risk-free routes, and recognize and avoid potential hazards in order to function effectively in dynamic and unpredictable environments. While there has been a lot of development in the field of autonomous navigation, the majority of current methods still rely on simplified models and assumptions that may not accurately reflect the complexity and variability of the environments in which autonomous vehicles operate. New methods of autonomous navigation based on environmental learning are being developed to give robots and other autonomous systems the ability to learn from their surroundings. Environmental learning is an application of machine learning and artificial intelligence that equips robots with the ability to perceive and understand their surroundings, adapt to new situations, and gradually improve their performance based on what they've learned [1, 2].

In this chapter, a high-level look at environmental learning for autonomous navigation is provided, exploring its theoretical underpinnings, practical applications, and cutting-edge methods. First, an overview of autonomous navigation is presented, outlining its fundamental concepts and algorithms before delving into the difficulties of this task in real-world settings with a variety of obstacles. A broad introduction to environmental learning is discussed, covering its many varieties as well as the most recent developments in the field. Recent studies demonstrating the efficacy of environmental learning in a variety of applications are discussed, highlighting its potential for better path planning and obstacle avoidance [3–5].

13.1.1 ENVIRONMENT LEARNING

In order for vehicles and robots to function reliably and efficiently in high-variability settings, autonomous navigation will be an essential component of future transportation and robotics. Autonomous navigation relies heavily on environmental learning, which allows the system to learn about its immediate context and act accordingly. Mapping the environment, identifying and avoiding obstacles, selecting the optimal route to the destination, and making decisions in real time based on sensor data are all part of the environmental learning process for autonomous navigation. Acquiring data about one's immediate surroundings and representing it usefully is what "environmental mapping" is all about. This data could include the locations of things like buildings, roads, and landmarks. By compiling this data, a map of the surrounding area is created, which aids in navigation and avoiding potential hazards.

Autonomous navigation relies heavily on the ability to detect and avoid obstacles. The system relies on sensing components like cameras, lidar, and radar to get a clear picture of its surroundings and accomplish this goal. After gathering environmental data, the system must use logic to figure out how to navigate safely. This necessitates making decisions in real-time according to the present state of the ecosystem, which includes the position, velocity, and direction of the system, as well as the location and motion of obstacles [6, 7].

13.1.2 ENVIRONMENT LEARNING ALGORITHMS

One of the most important aspects of autonomous navigation is finding the most efficient route there. Algorithms designed for path planning take into account both the current location and the desired endpoint to devise the most efficient travel plan. This method takes into account the current location, velocity, and heading of the system, as well as the current and potential future states of obstacles, to determine a path that is both safe and productive.

Finally, autonomous navigation relies heavily on the ability to make decisions in the moment. One way to do this is with sensor fusion, which takes data from various sensors and uses it to make better decisions in real time. This ensures the system can navigate safely and efficiently by allowing it to quickly adapt to new circumstances [8, 9].

13.2 OVERVIEW OF ENVIRONMENT LEARNING ALGORITHMS

Thus, environmental learning is an essential part of autonomous navigation, allowing vehicles and robots to function efficiently and safely in high-variability settings. Both the technical and ethical difficulties of autonomous navigation are being studied in an effort to improve environmental learning algorithms' accuracy and reliability. The continued development of this field has the potential to completely alter the ways in which we move through and engage with the world [10].

1. **Autonomous Navigation:** It relies heavily on the popular technique of simultaneous localization and mapping (SLAM). Algorithms for simultaneous localization and mapping (SLAM) use collected sensor data to generate a map of the surrounding area and determine where the robot is in relation to that map. Successful applications of SLAM have been made in a variety of real-world settings, making it a staple of robotics research [11].
2. **Techniques Predicated on Deep Learning:** Convolutional neural networks (CNNs) and recurrent neural networks (RNNs) are just two examples of the deep learning strategies used by autonomous navigation systems to gain insight into their environments [2]. Such approaches allow robots to accomplish difficult tasks because they can decipher complex patterns and relationships in sensory input [12].
3. **Reinforcement Learning:** A machine learning technique called reinforcement learning teaches a robot how to perform in a given setting in order to maximize a reward signal. Reinforcement learning has been successfully applied to autonomous navigation in complex environments, where the reward signal is typically defined based on the robot's performance, such as how quickly it completes a task or how well it avoids obstacles [13].
4. **Bayesian Approaches:** Bayesian methods have been used to model the uncertainty in environmental data and estimate the robot's position and orientation. These approaches are particularly useful in scenarios where sensor data is noisy or unreliable [14].
5. **Hybrid Approaches:** This that combine multiple techniques, such as SLAM, deep learning, and reinforcement learning, have

been proposed for environmental learning for autonomous navigation. These approaches aim to leverage the strengths of different techniques to overcome their individual limitations [15].

6. **Semantic Segmentation:** It is a technique that involves assigning a label to each pixel in an image. This technique has been used to segment images captured by a robot's sensors into different categories, such as roads, sidewalks, buildings, and trees. By using this technique, the robot can create a more detailed map of its environment and plan its navigation accordingly [16].
7. **Point Cloud Processing:** It involves processing 3D point clouds generated by a robot's sensors, such as LIDAR or RGB-D cameras. This technique has been used to extract features, such as edges and corners, from the point cloud data, using which a topographical map of the area can be constructed. Point cloud processing has also been used for obstacle detection and avoidance, as well as for localization and mapping [17].
8. **Transfer Learning:** One definition of "transfer learning" is the process of applying skills and knowledge gained in one context to another. This method has been applied to autonomous navigation's environmental learning to facilitate knowledge transfer between different environments. For example, a robot could be trained in a simulation of one environment and then use the knowledge learned in that simulation to navigate in a different environment [18].
9. **Multi-Modal Sensor Fusion:** The term "multi-modal sensor fusion" refers to the process of merging the data of many sensors to form a more accurate picture of the surrounding environment. This technique has been used to create more robust maps and to improve the accuracy of localization and mapping [19].
10. **Graphical Models:** These are probabilistic models that represent the relationships between variables using a graph. To estimate the robot's position and orientation and model the uncertainty in sensory input, this method has been employed in environmental learning for autonomous navigation. Graphical models have also been used to represent the relationships between different landmarks in the environment, which can be used for localization and mapping [20].

11. **3D Reconstruction:** These techniques involve constructing a 3D model of the environment from 2D images or point clouds. In autonomous navigation, these methods have been utilized for environmental learning to build a more precise map of the surrounding area. They can also be used to detect and avoid obstacles, as well as to assist in localization and mapping [21].
12. **Active Learning:** It is a machine learning technique that involves selecting the most informative data points to label. This method has been utilized to lessen the need for labeled data during model training in environmental learning for autonomous navigation. By selecting the most informative data points to label, the robot can learn more efficiently and with less data [22].
13. **Visual Odometry:** It is a technique that involves using visual features, such as edges and corners, to estimate the robot's motion. It has been used in autonomous navigation for environmental learning to approximate the robot's position and orientation with respect to the starting point. Combining visual odometry with other methods, like simultaneous localization and mapping (SLAM), is a common way to boost the precision of these processes [23, 24].
14. **Probabilistic Robotics:** It involves using probabilistic models to represent the uncertainty in sensory data and the robot's state. Environmental learning for autonomous navigation has made use of this method to estimate the robot's position and orientation, as well as to plan its navigation route. Probabilistic robotics has also been used to detect and avoid obstacles, and to create more accurate maps of the environment [25].
15. **Hybrid Geometric-Probabilistic Methods:** These involve combining geometric methods, such as point cloud processing, with probabilistic methods, such as Bayesian filtering. This technique has been used in environmental learning for autonomous navigation to create more accurate maps of the environment and to improve the accuracy of localization and mapping. Hybrid geometric-probabilistic methods have also been used for obstacle detection and avoidance [26].
16. **Deep Reinforcement Learning:** As a subfield of machine learning, deep reinforcement learning entails teaching a network to optimize its actions in light of past successes and failures. This

technique has been used in environmental learning for autonomous navigation to learn optimal navigation policies in complex and dynamic environments. Deep reinforcement learning has shown promising results in learning obstacle avoidance and path planning in challenging environments [27].

17. **Meta-Learning:** It is a type of machine learning that involves learning to learn. This technique has been used in environmental learning for autonomous navigation to improve the ability of robots to learn from new and unseen environments. By learning how to learn, robots can quickly adapt to new environments and learn more efficiently [28].
18. **Domain Adaptation:** The process of translating a machine learning model from one domain to another is known as domain adaptation. This technique has been used in environmental learning for autonomous navigation to adapt a model trained in one environment to another environment with different characteristics. By adapting the model to the new environment, the robot can navigate more effectively and with less training data [29].
19. **Generative Models:** Data similar to the training data can be generated using generative models, a form of machine learning model. This technique has been used in environmental learning for autonomous navigation to generate synthetic data that can be used to train models in simulation. By using synthetic data, robots can be trained more efficiently and effectively, as well as adapt to new environments more easily [30].

13.3 SIMULTANEOUS LOCALIZATION AND MAPPING (SLAM) [32, 33]

In robotics and computer vision, the technique of Simultaneous Localization and Mapping (SLAM) is used to map an environment while simultaneously estimating the robot's location within the map.

Using data collected from the robot's cameras and laser range finders, simultaneous localization, and mapping (SLAM) attempts to determine the robot's location and orientation within a scene. "Simultaneous Localization and Mapping" means that both the map-building and the robot's estimated position are done at the same time.

Bayesian filtering, and in particular the recursive Bayes filter, forms the mathematical foundation of SLAM. To infer the state of a system from imperfect data, the recursive Bayes filter can be used. The state of a SLAM system consists of the robot's pose (position and orientation) and a map of the environment. The map of the surrounding area makes up the system's state.

13.3.1 THE BASIC STEPS IN A SLAM ALGORITHM

1. **Prediction:** The robot's pose is predicted based on its previous pose and the control inputs, such as velocity or rotation, provided by the robot.
2. **Measurement:** The robot takes measurements using its sensors, such as a camera or lidar. These measurements are typically noisy due to sensor errors and environmental factors.
3. **Data Association:** The map features that correspond to the measurements have been linked to them. Specifically, this requires correlating information from the measurement with that from the map.
4. **Correction:** Measurements are linked to map characteristics in order to adjust the robot's position. Bayes' rule is used to adjust the robot's position and the current map.
5. **Map Update:** The map is updated based on the corrected robot's pose and the associated measurements.

Measurements are linked to map characteristics in order to adjust the robot's position. Bayes' rule is used to adjust the robot's position and the current map. The SLAM algorithm iteratively predicts, measures, associates, corrects, and updates the robot's pose and the map based on this probability distribution, allowing for accurate mapping and localization in real-world environments.

13.3.2 FORMULATION OF SLAM [34] USING A RECURSIVE BAYESIAN FILTER APPROACH

The algorithm iteratively predicts the robot's pose, updates it based on measurements, updates the map, and repeats the process at each time step. In practice, the PDFs are often approximated using techniques such as particle filters or Kalman filters. Moreover, the equations have been programmed into the robot's central processing unit.

1. **State Vector:** The state vector at time t, x(t), includes the robot's pose, represented by a vector [x(t), y(t), θ(t)], as well as the map of the environment, represented by a vector m(t).
2. **Motion Model:** Based on the prior stance, x_{t-1} and the control input u_t, the motion model estimates the robot's pose at time t. The formula is: $(x_t \mid x_{t-1}, u_t)$, where; 'p' is the probability density function (PDF) of the robot's motion: $p(x_t \mid x_{t-1}, u_t)$.
3. **Measurement Model:** The robot's position at any given moment, t, is related to the measurements taken at that time, z(t), according to the measurement model. The probability density function (PDF) of the measurements, given the robot's pose and the map, is given by: $p(z_t \mid x_t, m_t)$
4. **Bayes' Rule:** With a given set of measurements and control inputs up to time t, the posterior probability density of the robot's pose and the map is calculated using Bayes' rule as follows: $p(x_t, m_t \mid z_{1:t}, u_{1:t})$ where measurements and control inputs up to time t are represented by $z_{1:t}$ and $u_{1:t}$, respectively.

13.3.3 PREDICTION STEP

The prediction step involves computing the predicted state distribution, $bel^{-}(x_t)$, by integrating the motion model over the previous state distribution, $bel^{-}(x_{t-1})$, and the control input, u_t:$bel(x_t) = \int p(x_t \mid x_{t-1}, u_t)\, bel^{-}(x_{t-1})dx_{t-1}$

13.3.4 CORRECTION STEP

The correction step involves updating the predicted state distribution using the measurement model, $p(z_t \mid x_t, m_t)$, and the measurement, $z(t)$: $bel(x_t) = \eta p(z_t \mid x_t, m_t)\, bel^{-}(x_t)$; where \eta is a normalization constant that ensures the distribution integrates to one.

13.3.5 MAP UPDATE

The map is updated based on the corrected state distribution:

m_t=arg arg mqx $p(z_{1:t}, u_{1:t}, x_t)$ where $p(m' \mid z_{1:t}, u_{1:t}, x_t)$ is the posterior probability density of the map given the measurements, control inputs, and the robot's pose.

13.3.6 DATASETS USE TO WORK ON SIMULTANEOUS LOCALIZATION AND MAPPING (SLAM) [35]

These datasets are widely used in the robotics and computer vision communities, and there are many research papers and tutorials that use them as examples for SLAM algorithms. I hope this helps you get started with your SLAM project!

1. **The KITTI Vision Benchmark Suite:** This dataset includes a large collection of images and Velodyne point cloud data captured from a car driving through urban and suburban environments. It also includes ground-truth trajectories and camera calibration data.
2. **The Oxford RobotCar Dataset:** This dataset includes images, LiDAR, and GPS data captured by a car driving around Oxford, UK. It contains ground-truth trajectories and camera calibration data as well.
3. **The EuRoC MAV Dataset:** This dataset includes data from micro-aerial vehicles (MAVs) flying indoors and outdoors. It includes stereo camera images, IMU measurements, and ground-truth trajectories.
4. **The TUM RGB-D SLAM Dataset:** This dataset includes RGB-D images captured by a Microsoft Kinect sensor, as well as ground-truth trajectories. It is intended for use with SLAM algorithms that incorporate RGB-D data.

13.3.7 APPLICATIONS OF SLAM [20]

1. **SLAM can be used with a wide range of sensors:** While cameras and lidars are commonly used in SLAM, it can also be used with other types of sensors, such as GPS, magnetometers, and sonar.

This flexibility allows SLAM to be used in a wide range of environments, from indoors to outdoors, and from terrestrial to aerial robots.

2 **SLAM can be used in real-time applications:** Many SLAM algorithms are designed to work in real-time, which is essential for applications such as autonomous vehicles and drones. Real-time SLAM requires fast and efficient algorithms, as well as sensors that can capture data quickly and accurately.

3. **SLAM can be used for object recognition and tracking:** It is possible to employ SLAM algorithms for environmental item recognition and tracking, which is useful for applications such as robot grasping and manipulation. By using SLAM to map the world and monitor moving things within that map, robots can better understand and interact with their surroundings.

4. **SLAM is an active area of research:** SLAM is a challenging problem that requires the integration of many different technologies, such as computer vision, robotics, and optimization. As a result, there are several ongoing projects in this field of study to develop more accurate, efficient, and robust SLAM algorithms. New sensors, such as solid-state lidars, and new algorithms, such as deep learning-based approaches, are also being developed to improve SLAM performance.

5. **Multi-robot SLAM:** One area of research is the development of SLAM algorithms for teams of robots that can work together to map and navigate complex environments. Multi-robot SLAM is a challenging problem that requires coordination and communication between the robots, as well as the ability to fuse data from multiple sources.

6. **3D SLAM:** While many SLAM algorithms are designed for 2D environments, there is growing interest in developing algorithms that can create 3D maps of environments. 3D SLAM requires more complex sensors, such as 3D lidars, and more advanced algorithms that can handle the increased complexity of 3D data.

7. **SLAM for augmented and virtual reality:** Accurate 3D models of physical spaces can be generated using SLAM technology and

put to use in AR/VR and other context-aware technologies. A 3D model of a room built with SLAM might be used to insert virtual things in the real world and vice versa.

8. **SLAM for human-robot interaction:** SLAM can also be used to help robots better understand and interact with humans in shared environments. Robots can better predict and respond to human actions when they create a map of the environment and track the locations and movements of people.

13.3.8 FUTURE INNOVATIVE AREAS OF SLAM TECHNOLOGY

Future innovative areas of SLAM technology [37] is being used in a wide range of applications, and new use cases are emerging all the time. Here are a few examples of new and innovative applications of SLAM:

1. **Construction and Maintenance:** SLAM can be used to create 3D maps of construction sites and buildings, which can help with tasks such as progress tracking, quality assurance, and safety monitoring. SLAM can also be used for maintenance tasks, such as inspecting and repairing pipelines and other infrastructure.
2. **Agriculture:** SLAM can be used to create maps of agricultural fields and monitor crop growth, which can help with tasks such as yield estimation, disease detection, and irrigation management. SLAM can also be used to guide autonomous agricultural vehicles, such as tractors and harvesters.
3. **Entertainment:** SLAM can be used to create immersive experiences in entertainment applications, such as video games and theme park attractions. By creating 3D maps of real-world environments and tracking the movements of users, SLAM can create more realistic and interactive experiences.
4. **Healthcare:** SLAM can be used to create maps of hospital environments and track the movements of patients and staff, which can help with tasks such as resource management, infection control, and patient monitoring. SLAM can also be used in rehabilitation applications, such as guiding patients through physical therapy exercises.

5. **Manufacturing:** SLAM can be used to optimize manufacturing processes by creating maps of production environments and tracking the movement of goods and workers. SLAM can also be used for quality control tasks, such as inspecting and sorting products on assembly lines.

These are just a few examples of the many new and innovative applications of SLAM that are emerging. As the technology continues to improve, it is likely that even more new use cases will be developed.

13.4 ENVIRONMENT-LEARNING STRATEGY BASED ON DEEP LEARNING [12, 39]

In recent years, deep learning has emerged as a useful method for autonomous navigation environment learning. Here are some recommendations for creating an environment-learning strategy based on deep learning:

- To sum up, creating a deep learning-based method for environment learning is an involved and iterative process. However, reliable and efficient models for autonomous navigation can be developed with the proper information, resources, and methods.
- First, sensors like cameras, LIDAR, and RADAR can provide data for autonomous vehicle navigation. The information is typically displayed as a tensor or image with many channels, where each channel stands for a unique sensor input.
- Second, an image recognition and processing architecture based on convolutional neural networks (CNNs). Most commonly, a CNN will have several convolutional layers followed by one or more fully connected layers. Features are extracted from the input data via filters in the convolutional layers, and the fully connected layers then use these features to make predictions.
- Filters associated with each convolutional layer in a CNN are called convolutional filters, and they are used to sift through the input data and identify relevant features. During training, the weight matrices that stand in for the filters are tweaked to reduce the gap between the predicted and observed outputs.
- Fourth, "pooling" is a down sampling operation typically used in CNNs to make their feature maps smaller and introduce spatial invariance.

- After the input is transformed into a nonlinear output, activation functions are applied to the output of each neuron in the CNN. Some well-liked activation functions include the hyperbolic tangent function and the rectified linear unit (ReLU) function.
- Mean squared error (MSE) and cross-entropy loss are two examples of regression and classification loss functions, respectively, that can be used for autonomous vehicle navigation. The performance of the model is evaluated by the loss function, which computes the absolute value of the difference between the predicted and observed output.

In order to minimize the loss function, the CNN's weights and biases are updated using backpropagation, which brings us to step seven. Adjustments to the weights and biases are made to minimize the loss function by computing their derivative with respect to the current values of the weights and biases, which is done during backpropagation. Stochastic gradient descent (SGD), Adam, and Adagrad are just a few examples of optimization algorithms that can be used to fine-tune the CNN's weights and biases during training. Using a convolutional neural network (CNN) to extract features from the input data, followed by one or more fully connected layers to make predictions, constitutes the mathematical framework for environment learning in autonomous vehicle navigation using deep learning. Backpropagation is used in conjunction with an optimization algorithm to minimize the loss function and update the CNN's weights and biases. The CNN's performance can be optimized by adjusting its architecture and hyperparameters.

13.4.1 MATHEMATICAL FORMALIZATION OF THE PROBLEM OF ENVIRONMENT LEARNING FOR AUTONOMOUS VEHICLE NAVIGATION [40]

Let x be the set of input data, where each data point $x_i \epsilon x$ represents a multi-channel image or tensor obtained from the sensors on the autonomous vehicle. Let Y be the set of desired outputs, where each output $y_i \epsilon Y$ represents the desired action or control signal for the vehicle to navigate in the environment. In order to achieve the desired output y_i, it is necessary to learn a function $y_i(x_i)$ that maps the input data x_i. The neural network representation of this function looks like this: $y_i = f(x_i;\theta)$ with θ parameters.

Each of the L layers in the network is made up of neurons that add up their input values and run them through an activation function. Each layer's weights and biases are stored in matrices $Wl^{(l)}$ and bias vectors $b^{(l)}$), respectively. The output of each layer l is computed as follows: $z^{(l)}= W^{(l)}a^{(l-1)}+b^{(l)}$ $a^{(l)}=\sigma(z^{(l)})$ where σ is an activation function, such as the rectified linear unit (ReLU) or sigmoid function, and $a^{(l-1)}$ is the output of the previous layer. The input to the first layer is the input data x_i, such that $a^{(0)}= x_i$. The output of the final layer is the predicted output of the network, such that: $\hat{y}i=a^{(L)}$.

The network is trained by minimizing a loss function $L(\hat{y}_i,y_i)$ that measures the difference between the predicted output $\hat{y}i$ *an* d the desired output *yi*. The loss function is typically a regression loss, such as mean squared error (MSE), or a classification loss, such as cross-entropy loss. The parameters θ of the network are updated during training using an optimization algorithm such as stochastic gradient descent (SGD), such that:

$$\theta = \theta - a\partial L(\hat{y}_i,y_i) / \partial\theta$$

where; α is the learning rate; and $\partial L(\hat{y}_i,y_i) / \partial\theta$ is the gradient of the loss function with respect to the network parameters θ. The gradient is computed using backpropagation, which is a method for efficiently computing the gradients of the loss function with respect to the weights and biases of the network.

The problem of environment learning for autonomous navigation can be formulated as learning a function $f(x_i)$ that maps input data x_i to the desired output *yi* using a neural network and optimizing the network parameters θ to minimize the loss function $L(\hat{y}_i,y_i)$. The network architecture, activation functions, loss function, and optimization algorithm can be chosen and tuned to achieve the best performance for the specific task at hand.

13.5 REINFORCEMENT LEARNING ALGORITHM FOR ENVIRONMENT BY AUTONOMOUS VEHICLE NAVIGATION [13, 41]

To train an agent to operate in a given environment in a way that maximizes a reward signal, we use a machine learning technique called reinforcement learning. To clarify, the agent in the context of autonomous vehicle navigation is the vehicle itself, while the environment is the physical world the vehicle is navigating through. Reinforcement learning provides a powerful framework for autonomous vehicle navigation, as it allows the vehicle to learn how to navigate in complex, dynamic environments by interacting with the environment and optimizing its behavior based on a reward signal. However, reinforcement learning can be computationally expensive and requires careful tuning of hyperparameters and algorithms to achieve good performance.

A mathematical formulation of the reinforcement learning framework for environment learning in autonomous vehicle navigation:

1. **State Space:** Let S be the state space, where $s \in S$ represents the state of the environment at a given time step.
2. **Action Space:** Let A be the action space, where $a \in A$ represents the action taken by the autonomous vehicle at a given time step.
3. **Reward Function:** Let $R(s, a)$ be the reward function, which maps a state-action pair to a numerical reward signal. The goal of the vehicle is to learn a policy that maximizes the cumulative reward over a long-term horizon.
4. **Policy:** Let $\pi(a \mid s)$ be the policy, which is a conditional probability distribution over actions given a state. The policy is typically represented as a neural network that takes the current state as input and outputs a probability distribution over possible action.
5. **Value Function:** Let $V(s)$ be the value function, which estimates the expected cumulative reward from a given state. The value function is used to evaluate the quality of a given policy and can be learned through iterative updates using the Bellman equation.

The reinforcement learning process involves the vehicle interacting with the environment, observing the state and reward signals, and updating its policy and value function based on the observed data. Specifically, the process consists of the following steps:

i. At time t, the vehicle observes the current state s_t of the environment and selects an action a_t based on its current policy: $a_t \sim \pi(a|s_t)$.
ii. The environment transitions to a new state s_{t+1} based on the selected action a_t, and the vehicle receives a reward signal $r_t = R(s_t, a_t$.
iii. The vehicle updates its value function based on the observed state, action, and reward using the Bellman equation:

$$V(s_t) \leftarrow V(s_t) + a[r_t + \gamma V(s_{t+1}) - V(s_t)]$$

where; α is the learning rate; and γ is the discount factor, which determines the importance of future rewards.

vi. The vehicle updates its policy based on the updated value function, using a policy optimization algorithm such as policy gradient or Q-learning.

v. Steps i–iv are repeated for many iterations, allowing the vehicle to gradually learn a better policy that maximizes the expected cumulative reward.

The reinforcement learning framework provides a powerful mathematical framework for autonomous vehicle navigation, allowing the vehicle to learn how to navigate in complex, dynamic environments by interacting with the environment and optimizing its behavior based on a reward signal.

13.6 BAYESIAN APPROACHES FOR ENVIRONMENT LEARNING BY AUTONOMOUS VEHICLE NAVIGATION [41]

Bayesian approaches provide a powerful framework for learning and reasoning about uncertainty in a wide range of applications, including environment learning for autonomous vehicle navigation. Bayesian methods allow for the incorporation of prior knowledge or beliefs about the environment into the learning process, which can help to improve the accuracy and reliability of the learned model. A Bayesian approach to environment learning for autonomous vehicle navigation can be formulated as follows:

1. **State Space:** Consider S as the state space, where $s \in S$ denotes the environment's state at a specific time step.
2. **Observation Space:** Let O be the observation space, where $o \in O$ represents the observation made by the autonomous vehicle at a given time step.
3. **Prior Distribution:** Let P(s) be the prior distribution over the state space, which represents the vehicle's prior beliefs about the environment before any observations are made.
4. **Likelihood Function:** Let $P(o|s)$ be the likelihood function, which represents the probability of observing o given the current state s.

5. **Posterior Distribution:** Let P(s|o) be the posterior distribution over the state space, which represents the vehicle's updated beliefs about the environment after making an observation o.
6. **Policy:** Let π(a|s) be the policy, which is a conditional probability distribution over actions given a state. The policy is usually depicted as a neural network that takes the current state as input and generates a probability distribution over possible action.

   The Bayesian approach to environment learning for autonomous vehicle navigation involves updating the posterior distribution over the state space as new observations are made and using this distribution to inform the policy. Specifically, the process consists of the following steps:

i. At time *t,* the vehicle observes the environment and makes an observation $o_t \in o$.
ii. The vehicle updates its posterior distribution over the state space using Bayes' rule:

$$P(o_{1:t}) = \frac{P(s)P(o_{1:t-1})}{P(o_{1:t-1})}$$

   where; $p(o_{1:t})$ represents the posterior distribution over the state space given all the observations up to time *t*; and $p(o_{1:t-1})$ is the marginal likelihood or evidence.
iii. The vehicle updates its policy based on the current posterior distribution over the state space, using a policy optimization algorithm such as Monte Carlo Tree Search (MCTS) or Bayesian optimization.

Steps i–iii are repeated for many iterations, allowing the vehicle to gradually improve its understanding of the environment and learn a better policy.

The Bayesian approach provides a powerful mathematical framework for environment learning in autonomous vehicle navigation, allowing the vehicle to reason about uncertainty and incorporate prior knowledge or beliefs about the environment into the learning process. However, Bayesian approaches can be computationally expensive and require careful tuning of hyperparameters and algorithms to achieve good performance.

13.7 SEMANTIC SEGMENTATION FOR ENVIRONMENT LEARNING BY AUTONOMOUS VEHICLE NAVIGATION [44]

Semantic segmentation's goal in the context of autonomous vehicle navigation is to help the vehicle recognize and categorize the various elements of its surroundings so that it can act appropriately in its navigation. By dividing the environment into distinct semantic regions, the car can more accurately detect road boundaries, steer clear of hazards, and locate points of interest.

As a rule, the semantic segmentation model is taught using huge collections of annotated still images or video stills depicting different driving scenarios. Different lighting, weather, and traffic densities can all be included in the dataset to test the model's adaptability and generalizability. As soon as the model has been trained, it can be incorporated into the autonomous vehicle's navigation system, where it will receive input from various sensors, including cameras, lidars, and radars, in the form of still images and video frames. The scene's various components—including roads, pedestrians, buildings, and vehicles—are all marked out on a segmentation map generated by the semantic segmentation model.

Semantic segmentation, which helps the vehicle perceive and understand its environment, is thus an essential part of autonomous vehicle navigation, allowing for safer and more efficient navigation. Partitioning an image into distinct regions or segments according to its semantic content (i.e., labeling each pixel in the image with a specific object or class) is a common computer vision task known as semantic segmentation. For autonomous vehicles, semantic segmentation is crucial because it enables them to learn about and navigate their surroundings by classifying the various objects they encounter. Training a deep neural network to recognize and classify different objects in the environment, like roads, pedestrians, traffic signs, buildings, and trees, is a crucial part of semantic segmentation for environment learning by autonomous vehicle navigation. Sensor data from cameras, lidars, and radars are fed into the neural network, and it outputs a segmentation map that labels the various objects it detects.

As soon as the semantic segmentation model has been trained, it can be incorporated into the autonomous vehicle's navigation system. The segmentation map is used by the navigation system to make calculated decisions about the vehicle's path, such as where to drive, how to navigate around obstacles, and where to find conveniences like parking and charging stations. Accurate and thorough labeling of the training data is required to make semantic segmentation a feasible task. Several factors, including the structure of the neural network and the optimization algorithm used during training, contribute to the final accuracy of the model.

13.7.1 MATHEMATICAL FORMULATION [16]

Semantic segmentation can be formulated mathematically as follows:

Let X be an input image of size W x H x C, where W, H, and C represent the width, height, and number of channels of the image, respectively. Each pixel in the image is represented as $x(i,j)$, where i and j denote the pixel's row and column indices. Let Y be the output segmentation map of size W x H x K, where K is the number of semantic classes. Each pixel in the segmentation map is represented as $Y(i, j, k)$, where k represents the class label assigned to the pixel.

The goal of semantic segmentation is to learn a function $f(X; \theta)$ that maps an input image X to its corresponding segmentation map Y, where θ represents the learnable parameters of the neural network. The function $f(X; \theta)$ can be represented as a convolutional neural network (CNN), which takes X as input and produces Y as output. The CNN consists of multiple convolutional layers, followed by non-linear activation functions and pooling layers, which help to extract meaningful features from the input image. The training process involves optimizing the parameters θ of the CNN to minimize a loss function $L(Y; \hat{Y})$, where $\hat{Y}$ is the predicted segmentation map produced by the CNN. The loss function measures the difference between the predicted segmentation map and the ground truth segmentation map.

13.7.1.1 ARCHITECTURE

An encoder-decoder network with skip connections is a common component of semantic segmentation models used in autonomous vehicle navigation. Features are extracted from the input image by the encoder network, which uses convolutional layers of progressively higher spatial resolution and lower channel dimensions to do so. The decoder network takes the encoder's feature maps and up samples them using transposed convolutional layers to create a segmentation map with the same spatial dimensions as the input image. In order to enhance the precision of the segmentation, skip connections are used to merge the high-resolution feature maps produced by the encoder with the up sampled feature maps produced by the decoder.

Let $f(X; \theta)$ be the semantic segmentation model with parameters θ, which takes an input image X of dimensions W x H x C and produces a segmentation map Y of dimensions W x H x K, where K is the number of semantic classes. The encoder network can be represented as a function $g(X; \phi)$, where ϕ are the encoder parameters, and the decoder network can be represented as a function $h(F; \psi)$, where F is the concatenated feature maps from the encoder, and ψ are the decoder parameters. The segmentation map Y can be expressed as:

$$Y = h\,(g\,(X; \phi); \psi)$$

13.7.1.2 TRAINING

The semantic segmentation model can be trained using a dataset of labeled images or video frames. The dataset can be split into training, validation, and test sets, with the training set used to update the model parameters, and the validation set used to monitor the model's performance and prevent overfitting.

The loss function used to train the model is typically the pixel-wise cross-entropy loss, defined as:

$$L(Y,\hat{Y}) = -\Sigma_{i,\,j,\,k}\, Y(i, j, k)\ log(\hat{Y}(i, j, k)\ Y(i, j, k)\ log(Y^\wedge(i, j, k)$$

where; $Y(i, j, k)$ is the ground truth label of *pixel* (i, j) for *class* k, and $\hat{Y}$ (i, j, k) is the predicted label for the same pixel and class.

13.7.1.3 THE OPTIMIZATION PROBLEM FOR SEGMENTATION

The optimization problem for segmentation can be formulated as follows:

minimize $\theta\ L(Y, \hat{Y})$ subject to:

$$\hat{Y} = f\,(x;\theta)$$

where; $f\,(X; \theta)$ is the CNN model, and Y is the ground truth segmentation map.

The optimization problem can be solved using gradient descent-based optimization algorithms such as stochastic gradient descent (SGD) or Adam.

The parameters θ of the CNN are updated iteratively by computing the gradients of the loss function with respect to θ and adjusting the parameters in the opposite direction of the gradients. With the backpropagation algorithm, we can train the model by computing gradients of the loss function with respect to the parameters θ, and then updating the parameters in the direction that minimizes the loss. Stochastic gradient descent (SGD) or a variant of SGD, like Adam, can be used as the optimization algorithm during model training.

Integrating a trained semantic segmentation model into an autonomous vehicle's navigation system allows for real-time processing of input images and generation of segmentation maps. Several tasks, including lane detection, object detection, and obstacle avoidance, are possible thanks to the segmentation map, all of which are critical to the effective and secure navigation of autonomous vehicles.

13.7.1.4 VALIDATION

Validating the semantic segmentation model's performance in a variety of real-world scenarios is crucial for ensuring the model's accuracy and reliability. This can be accomplished through both theoretical and practical testing of autonomous vehicles outfitted with the segmentation model on datasets that represent various driving scenarios. Metrics like intersection over union (IoU) and mean average precision can be used to assess the semantic segmentation model's efficacy on a test set (mAP). Similar to how mAP measures the average precision of object detection, IoU measures the overlap between the predicted and ground truth segmentation maps for each class.

Briefly, an encoder-decoder network with skip connections, fed a dataset of labeled images or video frames, and validated with metrics like IoU and mAP can serve as the basis for a semantic segmentation model for autonomous vehicle navigation. Improvements in model performance and generalization can be achieved through the use of regularization techniques such as weight decay, dropout, and data augmentation, as well as parameter updates using SGD or one of its variants.

All things considered, semantic segmentation is an essential part of autonomous vehicle navigation because it allows the vehicle to perceive and understand its surroundings and make calculated decisions about how to move through them. A more sustainable and connected future is possible with the addition of semantic segmentation to the navigation system, which can greatly increase the safety and efficiency of autonomous driving.

13.8 TRANSFER LEARNING FOR AUTONOMOUS NAVIGATION VEHICLE [46, 47]

The basic premise of transfer learning is to apply what is learned and modeled in one task to another, where it can be used to better perform that second task. While it's true that no two tasks are the same, the premise of transfer learning is that the knowledge and models learned from one task can be useful for learning a related task.

The basic idea is that rather than starting completely from scratch, a previously trained model can be used as the basis for learning the new task. A pre-trained model can be used to perform a new task because it has already learned features that are important for the original task. It's up to the original task's designer to decide how closely the new task is related to the old one. The main benefit of transfer learning is that it can drastically cut down on the time and effort needed to learn a new task by using previously acquired knowledge. If you use pre-trained models, you can learn a new task with much less data and computation than you would need to learn it from scratch. This is especially helpful in situations where there is a scarcity of data, data collection is prohibitively expensive, or computational resources are constrained. The central idea behind transfer learning is to efficiently and effectively apply the knowledge and models gained from one task to boost performance on a related but different task. Better performance can be attained with less data and computation, speeding up development across many machine learning applications. Machine learning models for environmental learning in robot or autonomous vehicle navigation can benefit from transfer learning in a number of important ways. Some examples are as follows:

1. **Object Detection:** This, which includes the identification and localization of objects in the environment like pedestrians, vehicles, and obstacles, is a crucial task for robot or autonomous vehicle navigation. Using pre-trained models that have learned to recognize common objects from large datasets like COCO and fine-tuning these models on specific environments or tasks relevant to robot or autonomous vehicle navigation is an effective way to increase the accuracy of object detection models through transfer learning.

2. **Semantic Segmentation:** Segmenting images into meaningful categories by assigning a semantic label to each pixel helps robots and autonomous vehicles navigate their environments more effectively. Semantic segmentation models can be made more accurate through the use of transfer learning by adapting and refining pre-trained models that have learned to segment various objects and regions using large datasets such as Cityscapes or Mapillary and then applying those models to the task of navigating a robot or autonomous vehicle.
3 **Reinforcement Learning:** This, which is widely used to teach robots and autonomous vehicles to navigate in challenging environments by optimizing a reward function based on the data they collect as they engage with those environments. Using pre-trained models that have learned to navigate in similar environments as a jumping-off point for training a new model on a similar task is an example of transfer learning, which can be used to improve the efficiency of reinforcement learning.
4. **Localization and Mapping:** Building a map of the environment and estimating the robot's position within that map constitutes localization and mapping, two essential tasks for robot or autonomous vehicle navigation. By taking advantage of pre-trained models that have learned to build maps from sensor data and fine-tuning these models on specific environments or tasks relevant to robot or autonomous vehicle navigation, transfer learning can improve the accuracy and efficiency of localization and mapping models.

Because of its versatility, transfer learning is a potent tool for enhancing the effectiveness of machine learning models for environmental learning in robot or autonomous vehicle navigation.

13.8.1 TRANSFER LEARNING FOR ENVIRONMENTAL LEARNING OF AUTONOMOUS VEHICLE NAVIGATION [48]

By reusing the information and models developed for one task to better perform another, transfer learning is a machine learning technique. Transfer learning can be used to enhance the effectiveness of machine learning models in the context of environmental learning for autonomous vehicle navigation by drawing on the strengths of previously trained models for the same or similar tasks.

When it comes to tasks like environmental learning for autonomous vehicle navigation, the basic idea behind transfer learning is that models trained on large datasets for a specific task, like image recognition or object detection, can be repurposed and adapted for related tasks. When compared to training a model from scratch, this approach can save a lot of time and effort while potentially yielding better results.

Using pre-trained object detection models, such as those trained on the COCO dataset, and then fine-tuning them on specific environments or tasks relevant to autonomous vehicle navigation is an example of using transfer learning for environmental learning in autonomous vehicle navigation. The model can learn to recognize objects important to autonomous vehicle navigation, such as traffic signs, pedestrians, and vehicles, with less data and in less time by using pre-trained models as a starting point.

Transfer learning has several benefits in the context of environmental learning for autonomous vehicle navigation, including:

1. **Faster Training:** Transfer learning can significantly reduce the time required to train machine learning models for environmental learning by leveraging pre-trained models.
2. **Better Performance:** Transfer learning can potentially lead to better performance by leveraging knowledge and models learned from related tasks and datasets.
3. **Improved Generalization:** Transfer learning can improve the ability of machine learning models to generalize to new environments and tasks by incorporating prior knowledge from related tasks.
4. **Reduced Data Requirements:** Transfer learning can reduce the amount of data required to train machine learning models by leveraging pre-trained models.
5. **Scalability:** Transfer learning can be used to scale up machine learning models to new environments and tasks by building on prior knowledge and models.

Transfer learning is a powerful technique for improving the performance and efficiency of machine learning models for environmental learning in autonomous vehicle navigation. Transfer learning involves leveraging knowledge and models learned from one task, typically referred to as the source task, to improve performance on a related but different task, referred to as the target task.

13.8.2 MATHEMATICAL FORMULATION OF TRANSFER LEARNING [48, 49]

Let X be the input data, Y be the output labels, and f be a function that maps inputs to outputs. Given a source task $T_s=(X_s, Y_s, f_s)$ and a target task $T_t=(X_t, Y_t, f_t)$, where X_s and X_t are the input data for the source and target tasks, respectively, Y_s and X_t are the output labels, and f_s and f_t are the corresponding mapping functions. In transfer learning, the goal is to learn a mapping function $f_t(X_t)$ for the target task T_t by leveraging knowledge and models learned from the source task T_s. This can be done by using the pre-trained model from the source task to initialize the model for the target task, and then fine-tuning the model using data from the target task.

Steps to be followed for the transfer learning process:

1. Pre-train a model on the source task T_s using input data X_s and output labels Y_s to learn a mapping function $f_S(X_S)$ that can generalize to new data.
2. Initialize the model for the target task T_t using the pre-trained model from the source task.
3. Fine-tune the model on the target task using input data X_t and output labels Y_t to learn a mapping function $f_t(X_t)$ that is specific to the target task.
4. Evaluate the performance of the fine-tuned model on the target task to determine its effectiveness.

By applying what was learned and modeled in the first task (called the "source task") to the second (called the "target task"), performance on the "target task" can be enhanced. The model can be fine-tuned with data from the target task in order to learn task-specific features, and then the features it learned on the source task can be transferred to the target task. In conclusion, transfer learning provides a robust framework for enhancing the effectiveness and robustness of machine learning models for environmental learning in autonomous vehicle navigation, thereby enhancing performance.

13.9 POINT CLOUD PROCESSING [50]

The term "point cloud processing" is used to describe the methods used to derive useful information from 3D point clouds. A point cloud is a collection of 3D points used to represent the shape and structure of an object or scene. They are typically gathered with the help of structured light cameras and LiDAR (light detection and ranging) sensors. Autonomous vehicle navigation, 3D modeling, robotics, and augmented reality are just some of the many uses for point cloud processing. Point cloud processing is used in autonomous vehicle navigation, for instance, to detect and estimate the locations and motions of various environmental features, including other vehicles, pedestrians, and buildings.

Point cloud processing is a complex and multi-step process that requires expertise in signal processing, computer vision, and machine learning. However, it is a powerful tool for extracting valuable information from 3D point cloud data and enabling advanced applications in a variety of fields. Point cloud processing has several advantages and benefits, including:

1. **Accurate Representation of 3D Environment:** Point clouds provide a highly accurate and detailed representation of the 3D environment, which can be used to support a wide range of applications such as autonomous vehicle navigation, 3D modeling, and robotics.
2. **Ability to Capture Large Amounts of Data:** Point clouds can capture large amounts of data in a relatively short amount of time, making them an efficient and effective way to acquire detailed 3D information about a scene or object.
3. **Versatility:** Point cloud processing techniques can be applied to a wide range of sensor data types, including LiDAR, structured light cameras, and stereo cameras, making it a versatile tool for extracting information from 3D data.
4. **Real-Time Processing:** Advances in point cloud processing algorithms and hardware have enabled real-time processing of point cloud data, making it possible to use point clouds in real-time applications such as autonomous vehicle navigation.

5. **Increased Efficiency:** Point cloud processing can help increase efficiency in various industries such as construction, architecture, and manufacturing. It can help streamline workflows, improve accuracy, and reduce costs.
6. **Improved Safety:** Point cloud processing can be used to identify potential hazards in a given environment, such as obstacles in the path of an autonomous vehicle or safety hazards in a construction site.

Point cloud processing is a powerful tool for extracting valuable information from 3D point cloud data and enabling advanced applications in a variety of fields. Its versatility, accuracy, and efficiency make it an attractive solution for many industries and applications.

13.9.1 CONSTRAINTS AND CHALLENGES THAT NEED TO BE CONSIDERED TO IMPLEMENT POINT CLOUD PROCESSING

1. **Data Size:** Point clouds can be very large and complex, which can create challenges in terms of storage, processing, and analysis. This requires powerful computing resources and specialized software to manage and process the data.
2. **Noise and Outliers:** Point cloud data can be affected by noise and outliers, which can make it difficult to extract meaningful information. Techniques such as filtering, and outlier removal must be used to address these issues.
3. **Segmentation Errors:** Segmentation algorithms may not always accurately separate objects in the point cloud, leading to errors in object recognition and classification.
4. **Limited View:** Point clouds are generated by sensors such as LiDAR or structured light cameras, which have limited viewing angles and may not capture all aspects of a scene or object. This can lead to incomplete or inaccurate representations of the environment.

5. **High Cost:** Point cloud sensors and processing software can be expensive, making it difficult for some industries to adopt point cloud processing technology.
6. **Time-Consuming:** Point cloud processing can be a time-consuming process, particularly for large datasets. This can limit its effectiveness for real-time applications or require extensive computing resources.

   Point cloud processing is a powerful tool for extracting valuable information from 3D point cloud data, but it also has constraints and challenges that must be carefully considered in the design and implementation of point cloud processing systems.

13.9.2 ALGORITHM: POINT CLOUD PROCESSING FOR ENVIRONMENTAL LEARNING OF AUTONOMOUS VEHICLE NAVIGATION [51]

Point cloud processing is an important technique for environmental learning in autonomous vehicle navigation. Point clouds are 3D data structures that represent the geometry and spatial information of objects in the environment. In the context of autonomous vehicles, point clouds are typically acquired using lidar sensors, which emit laser beams and measure the time of flight to determine the distance and position of objects in the environment. Point cloud processing involves several steps, including pre-processing, segmentation, feature extraction, and classification. These steps can be mathematically formulated as follows:

1. **Pre-Processing:** The raw point cloud data is pre-processed to remove noise, outliers, and other artifacts that can affect the accuracy of the subsequent processing steps. The pre-processing steps typically include filtering, down-sampling, and registration.

   Let $P = p_1, p_2, \ldots, p_n$ be the raw point cloud data, where $p_i=(x_i, y_i, z_i)$ is a 3D point in Cartesian coordinates. The pre-processing steps can be represented as a function $f(P; \alpha)$, where α is the pre-processing parameters. The pre-processed point cloud data can be represented as $P' = f(P; \alpha)$.

2. **Segmentation:** The pre-processed point cloud data is segmented into clusters that correspond to individual objects in the environment. Segmentation can be performed using various techniques such as region growing, clustering, and graph-based methods.

   Let $S = s_1, s_2, \ldots, s_k$ be the segmented point cloud data, where $s_i = p_i 1, p_i 2, \ldots, p_i m$ is a cluster of points corresponding to object i. The segmentation process can be represented as a function $g(P'; \beta)$, where β is the segmentation parameters. The segmented point cloud data can be represented as $S = g(P'; \beta)$.

3. **Feature Extraction:** Features are extracted from the segmented point cloud data to capture the shape, orientation, and other characteristics of the objects in the environment. Feature extraction can be performed using various techniques such as shape descriptors, surface normals, and histograms of local features.

   Let $F = f_1, f_2, \ldots, f_k$ be the feature vectors extracted from the segmented point cloud data, where fi is the feature vector for object i. The feature extraction process can be represented as a function $h(S; \gamma)$, where γ represents the feature extraction parameters. The feature vectors can be represented as $F = h(S; \gamma)$.

4. **Classification:** The feature vectors are used to classify the objects in the environment into different semantic classes such as cars, pedestrians, and buildings. Classification can be performed using various techniques such as machine learning algorithms and rule-based methods.

   `Let $C = c_1, c_2, \ldots, c_k$ be the semantic classes of the objects in the environment, where c_i is the semantic class of object i. The classification process can be represented as a function $k(F; \delta)$, where δ represents the classification parameters. The semantic classes can be represented as $C = k(F; \delta)$.

   In summary, point cloud processing for environmental learning in autonomous vehicle navigation involves pre-processing, segmentation, feature extraction, and classification steps, which can be mathematically formulated using various functions and parameters. These techniques enable autonomous vehicles to accurately perceive and navigate their environment, which is crucial for safe and efficient operation.

13.10 MULTI-MODAL SENSOR FUSION [45, 52]

For a more complete picture of the environment or object being sensed, it is common practice to fuse data from multiple sensors that use different modalities. Vision, sound, touch, and other senses can all be used as modalities. Multi-modal sensor fusion is used in autonomous vehicles, for instance, to compile information from various sensors (such as cameras, lidars, radars, and GPS) into a unified picture of the environment. Because of this, the car can make smart decisions about its path and avoid obstacles.

Robotics, medical diagnosis, and other fields that require a more complete picture of the world around or within an object are also utilizing multi-modal sensor fusion. Data from multiple sensors must be pre-processed, features extracted, and then combined using algorithms like Kalman filters or neural networks to form the fusion. For a more complete picture of the environment or object being sensed, it can be helpful to fuse data from multiple sensors that use different modalities. Specifically, this requires the following technical procedures:

1. **Pre-Processing:** The first step is to pre-process the data from each sensor to ensure that it is in a suitable format for fusion. This may involve correcting for sensor biases, scaling the data to a common range, or aligning the data in time or space.
2. **Feature Extraction:** Once the data has been pre-processed, the next step is to extract features from each sensor modality. These features may include things like edge detection in visual data, frequency analysis in audio data, or pattern recognition in haptic data.
3. **Alignment:** After feature extraction, the data from each sensor modality needs to be aligned in time and space. This ensures that the features extracted from each modality are synchronized and can be combined properly.
4. **Fusion:** The last step is to use an algorithm, like a Kalman filter, Bayesian network, or neural network, to combine the features

extracted from each sensor modality. In order to create a more complete and accurate representation of the environment or object being sensed, these algorithms take the features as inputs and combine them using statistical or machine learning techniques.

Dealing with data uncertainty and noise is a major obstacle to multi-modal sensor fusion. Errors introduced by combining data from different sensor modalities are amplified, as each modality may have its own sources of error. Robust algorithms that can deal with noisy data are needed, as well as careful modeling and estimation of uncertainty. Equally difficult is figuring out how to give each sensor modality its due in the fusion process. This is conditional on a number of variables, including the sensors' individual precision and accuracy, the nature of the work being done, and the surrounding conditions. Expertise in signal processing, statistics, and machine learning is required for the complex and difficult problem of multi-modal sensor fusion. If implemented properly, however, it can yield enormous gains in precision, sturdiness, and situational awareness across a wide variety of contexts. There are several approaches that can be taken to overcome the challenges of multi-modal sensor fusion:

1. **Sensor Calibration:** Accurate calibration of the sensors is crucial to reduce the bias and error in the data. This can be done by performing regular sensor calibrations and by using calibration models to correct for sensor biases.
2. **Noise Reduction Techniques:** There are various techniques to reduce the noise in the data from each sensor modality, such as filtering, smoothing, and denoising. These techniques can be applied before feature extraction or during the fusion process to reduce the effect of noise.
3. **Feature Selection:** Not all features extracted from each sensor modality may be relevant for the fusion process. Selecting the most informative features can reduce the noise and improve the accuracy of the fused data.
4. **Fusion Algorithms:** There are various fusion algorithms available that can be used to combine the data from multiple sensors. These algorithms differ in their assumptions, computational complexity, and accuracy. Choosing the appropriate fusion algorithm can have a significant impact on the accuracy and reliability of the fused data.

5. **Uncertainty Modeling:** Modeling and estimating the uncertainty in the data is important to reduce the effect of noise and errors in the fusion process. This can be done using Bayesian networks, Kalman filters, or other statistical methods that can account for uncertainty.
6. **Contextual Information:** Incorporating contextual information, such as the task being performed or the environment in which the sensors are operating, can improve the accuracy and robustness of the fused data.
7. **Machine Learning:** For optimal data fusion, it is possible to use machine learning methods to learn the connections between the data collected by various sensors. This may involve any of a variety of techniques, such as supervised learning, unsupervised learning, or reinforcement learning.

In conclusion, multi-modal sensor fusion problems can only be solved with a mix of technical knowledge, uncertainty modeling, and the right choice of fusion algorithms and techniques. As these issues are resolved, the quality of the fused data (as well as its reliability and utility) will increase dramatically. To help autonomous vehicles understand their surroundings and navigate safely, multi-modal sensor fusion is an essential technology. When multiple sensors report the same information, autonomous vehicles have a more complete picture of their surroundings and can make better decisions about where to go and what to do. Learning the environment through autonomous vehicle navigation is aided by multi-modal sensor fusion in a number of ways:

1. **Object Detection and Tracking:** By combining data from cameras, lidars, and radars, autonomous vehicles can detect and track objects such as other vehicles, pedestrians, and obstacles. This allows them to plan their movements and avoid collisions.
2. **Localization and Mapping:** Autonomous vehicles need to be able to localize themselves accurately within their environment and build maps of their surroundings. By combining data from GPS, inertial sensors, and lidars, they can create detailed maps of their environment and use this information to navigate.
3. **Environmental Sensing:** Autonomous vehicles need to be able to sense environmental conditions such as road surface quality,

weather conditions, and traffic patterns. By combining data from sensors such as cameras, lidars, and weather sensors, they can obtain a more accurate and comprehensive understanding of their environment.

4. **Path Planning and Decision-Making:** Autonomous vehicles need to be able to plan their paths and make decisions based on the data they have collected. By combining data from multiple sensors, they can make informed decisions about their movements, such as when to change lanes, when to slow down, or when to stop.

Multi-modal sensor fusion is a critical component of autonomous vehicle navigation, enabling vehicles to learn about their environment and make informed decisions based on the data they have collected. By using a combination of sensors and fusion algorithms, autonomous vehicles can achieve a high level of accuracy, reliability, and safety in their navigation.

13.10.1 MATHEMATICAL FORMULATION OF MULTI-MODAL SENSOR FUSION [43, 50]

Multi-modal sensor fusion can be modeled mathematically using a Bayesian framework, which allows for the integration of information from multiple sources and the representation of uncertainty. In this framework, the goal is to estimate the state of the environment based on the measurements obtained from multiple sensors.

Let x represent the state of the environment and let $z_1, z_2,...,z_n$ represent the measurements obtained from n different sensors. We can model the relationship between the state and the measurements using Bayes' theorem:

$P(x | z_1, z_2,...,z_n) = P(z_1, z2,...,z_n | x)P(x) / P(z_1, z_2,...,z_n)$ where $P(x | z_1, z_2,...,z_n)$ is the posterior probability distribution over the state given the measurements, $P(z_1, z2,...,z_n | x)$ is the likelihood function that describes the probability of obtaining the measurements given the state, P(x) is the prior probability distribution over the state, and $P(z_1, z2,...,z_n$ is the evidence or marginal likelihood of the measurements. To estimate the state of the environment, we need to compute the posterior probability distribution. This can be done using various fusion algorithms, such as Kalman filters, particle filters, or Markov Chain Monte Carlo (MCMC) methods. These algorithms use the likelihood function and the prior probability distribution to update the posterior distribution over time as new measurements are obtained.

The likelihood function can be modeled using the sensor models of each individual sensor. For example, a camera sensor model might describe the probability of observing an object at a certain location in the image given the state of the environment. Similarly, a lidar sensor model might describe the probability of observing a reflection at a certain angle and distance given the state of the environment. The prior probability distribution can be modeled using prior knowledge or assumptions about the state of the environment. For example, if we know that the vehicle is on a road, we might assume that the vehicle is within a certain distance of the center of the road.

In summary, multi-modal sensor fusion can be modeled mathematically using a Bayesian framework, which allows for the integration of information from multiple sources and the representation of uncertainty. The posterior probability distribution over the state of the environment can be estimated using fusion algorithms that update the distribution based on new measurements over time.

13.11 GENERATIVE MODELS FOR ENVIRONMENTAL LEARNING OF AUTONOMOUS NAVIGATION [30]

Machine learning techniques can be used to discover relationships between data from different sensors, allowing for more accurate fusion. This could be achieved through the use of a number of methods, including supervised learning, unsupervised learning, and reinforcement learning.

In conclusion, the difficulties associated with multi-modal sensor fusion can only be overcome by combining the appropriate application of technical expertise, uncertainty modeling, and fusion algorithms and techniques. After these problems are fixed, the fused data will be of much higher quality (not to mention reliability and utility). Multi-modal sensor fusion is a crucial technology for assisting autonomous vehicles in gaining situational awareness and safely navigating their environments. Autonomous vehicles have a more complete picture of their surroundings and can make more informed decisions when multiple sensors report the same information. Multi-modal sensor fusion can help autonomous vehicles in a number of ways when it comes to learning the environment through navigation.

13.11.1 AUTOENCODER: A GENERATIVE MODEL USED FOR ENVIRONMENTAL LEARNING [8]

When properly trained, neural networks can be used as autoencoders to reconstruct the original data. They are made up of an encoder network, which converts data to a lower-dimensional representation, and a decoder network, which converts the output back to the original data. Autoencoders can be trained using real sensor readings or images captured in a variety of environments, making them ideal for use in environmental learning for autonomous navigation. After being fed data, an autoencoder can learn to encode it into a representation with fewer dimensions while still preserving its essential characteristics. As a result, we can use the decoder network to generate synthetic data that closely resembles the real data but with slight modifications to reflect a variety of environmental conditions.

As with GANs, once the autoencoder has been trained, the synthetic sensor readings or images can be used to train the autonomous navigation system to function in a wide range of environments. In order to train an autonomous navigation system to function in a wide variety of environments, generative models can be an invaluable resource for environmental learning. Using a generative model (in this case, a GAN) for autonomous navigation environment learning with a simplified algorithm:

1. Collect a dataset of real sensor readings or images that represent different environmental conditions. For example, you might collect sensor readings or images from different weather patterns, lighting conditions, or terrain types.
2. Train a GAN on the dataset of real sensor readings or images. The GAN consists of a generator network and a discriminator network. The generator network takes a random noise vector as input and produces synthetic sensor readings or images, while the discriminator network tries to distinguish between the real and synthetic data.
3. During training, the generator network learns to produce synthetic sensor readings or images that are increasingly similar to the real ones, while the discriminator network learns to distinguish between the real and synthetic data.
4. Once the GAN has been trained, use the synthetic sensor readings or images to train the autonomous navigation system to operate in a variety of environmental conditions. For example, you might

use the synthetic data to train the system to recognize different types of terrain, or to navigate in low-light conditions.

5. During testing, the autonomous navigation system will encounter new environmental conditions that it has not seen before. However, because it has been trained on synthetic data that represents a variety of environmental conditions, it should be able to adapt to these new conditions and navigate successfully.

13.11.2 MATHEMATICAL FORMULATION OF GENERATIVE MODEL ALGORITHM [36, 38, 42]

A more formal mathematical formulation of the algorithm for using a generative model (specifically, a generative adversarial network or GAN) for environmental learning of autonomous navigation:

i. Collect a dataset of real sensor readings or images that represent different environmental conditions. Let X be the dataset, where each sample x_i is a vector or image representing an environmental condition.
ii. Train a GAN on the dataset of real sensor readings or images. The GAN consists of a generator network $(G(z;\theta_g)$ and a discriminator network $(D(x;\theta_d)$. G takes a random noise vector z as input and produces synthetic sensor readings or images $(G(z;\theta_g)$, while D tries to distinguish between the real data x and the synthetic data $(G(z;\theta_g)$
iii. During training, the generator network learns to produce synthetic sensor readings or images that are increasingly similar to the real ones, while the discriminator network learns to distinguish between the real and synthetic data. This is achieved by minimizing the following objective function:

$$min_G max_D \mathrm{E}_x Z\,[log(1-D(G(z;\theta_g);\theta_d))]$$

where; Z is the space of possible noise vectors z; and $\mathrm{E}_x Z$ denote the expected values over the real and synthetic data distributions, respectively.

iv. Once the GAN has been trained, use the synthetic sensor readings or images $(G (z;\theta_g)$ to train the autonomous navigation system to operate in a variety of environmental conditions. Let Y be the dataset of labels for the environmental conditions, where y_i is the label for x_i. Use the synthetic data $(G (z;\theta_g)$ and their corresponding labels y_i to train a classification or regression model $(M (x;\theta_m)$ that maps sensor readings or images x to actions or outputs.

v. During testing, the autonomous navigation system will encounter new environmental conditions that it has not seen before. However, because it has been trained on synthetic data that represents a variety of environmental conditions, it should be able to adapt to these new conditions and navigate successfully.

Note that there are many variations and extensions to this basic algorithm, and the details of the implementation can vary depending on the specific application and the choice of generative model.

KEYWORDS

- **adaptation**
- **autonomous navigation**
- **deep learning**
- **environmental learning**
- **reinforcement learning**
- **robot perception**
- **simultaneous localization and mapping (SLAM)**

REFERENCES

1. Badue, C., Guidolini, R., Carneiro, R. V., Azevedo, P., Cardoso, V. B., Forechi, A., Jesus, L., Berriel, R., Paixão, T. M., Mutz, F., et al., (2021). Self-driving cars: A survey. *Expert Syst. Appl., 165*, 113816.
2. Rosique, F., Navarro, P. J., Fernández, C., & Padilla, A., (2019). A systematic review of perception system and simulators for autonomous vehicles research. *Sensors, 19*(3), 648.

3. Kuutti, S., Fallah, S., Katsaros, K., Dianati, M., Mccullough, F., & Mouzakitis, A., (2018). A survey of the state-of-the-art localization techniques and their potentials for autonomous vehicle applications. *IEEE Internet Things*, *5*, 829–846.
4. Van, B. J., O'Brien, M., Gruyer, D., & Najjaran, H., (2018). Autonomous vehicle perception: The technology of today and tomorrow. *Transportation Research Part C: Emerging Technologies*, *89*, 384–406.
5. Liu, R., Wang, J., & Zhang, B., (2020). High-definition map for automated driving, overview and analysis. *J. Navig.*, *73*, 324–341.
6. Xiao, X., Liu, B., Warnell, G., Fink, J., & Stone, P. (2020). Appld: Adaptive planner parameter learning from demonstration. *IEEE Robotics and Automation Letters*, *5*(3), 4541–4547.
7. Finn, C., & Levine, S., (2017). Deep visual foresight for planning robot motion. In: *Robotics and Automation (ICRA), 2017 IEEE International*.
8. Danilo, B., Sylvain, C., & Darwin, G. C., (2016). Learning autonomous behaviors for the body of a flexible surgical robot. *Autonomous Robots*, 1–15.
9. Perille, D., Truong, A., Xiao, X., & Stone, P., (2020). Benchmarking metric ground navigation. In: *2020 IEEE International Symposium on Safety*.
10. Burgard, W., Stachniss, C., & Hähnel, D., (2007). *Autonomous Navigation in Dynamic Environments*. Berlin, Heidelberg: Springer-Verlag.
11. Zheng, S., Wang, J., Rizos, C., Ding, W., & El-Mowafy, A., (2023). Simultaneous localization and mapping (SLAM) for autonomous driving: Concept and analysis. *Remote Sens.*, *15*, 1156. https://doi.org/10.3390/rs15041156.
12. Hou, Y. L., Song, Y., Hao, X., Shen, Y., Qian, M., & Chen, H., (2018). Multispectral pedestrian detection based on deep convolutional neural networks. *Infrared Phys. Technol.*, *94*, 69–77.
13. Zhang, T., & Mo, H., (2021). Reinforcement learning for robot research: A comprehensive review and open issues. *International Journal of Advanced Robotic Systems*, *18*(3). doi: 10.1177/17298814211007305.
14. Ma, J., Xie, H., Song, K., & Liu, H., (2021). A Bayesian driver agent model for autonomous vehicles system based on knowledge-aware and real-time data. *Sensors (Basel)*, *21*(2), 331. doi: 10.3390/s21020331. PMID: 33418987; PMCID: PMC7825336.
15. Veronica, V., Gustavo, De. M. S., & Claudio, F. M. T., (2023). Harpia: A hybrid system for agricultural UAV missions. *Smart Agricultural Technology*, *4*, 100191, ISSN 2772-3755. https://doi.org/10.1016/j.atech.2023.100191.
16. Dubé, R., Cramariuc, A., Dugas, D., et al., (2020). SegMap: Segment-based mapping and localization using data-driven descriptors. *The International Journal of Robotics Research*, *39*(2, 3), 339–355. doi:10.1177/0278364919863090.
17. Dubé, R., Gollub, M. G., Sommer, H., et al., (2018b). Incremental segment-based localization in 3D point clouds. *IEEE Robotics and Automation Letters*, *3*(3), 1832–1839.
18. Kebria, P. M., et al., (2020). Autonomous navigation via deep imitation and transfer learning: A comparative study. In: *2020 IEEE International Conference on Systems, Man, and Cybernetics (SMC)* (pp. 2907–2912). Toronto, ON, Canada. doi: 10.1109/SMC42975.2020.9283065.

19. Melotti, G., Premebida, C., Gonçalves, N. M. D. S., Nunes, U. J., & Faria, D. R., (2018). Multimodal CNN pedestrian classification: A study on combining LIDAR and camera data. In: *Proceedings of the 2018 21st International Conference on Intelligent Transportation Systems (ITSC)* (pp. 3138–3143). Maui, HI, USA. IEEE: Maui, HI, USA.
20. Chang, L., Niu, X., Liu, T., Tang, J., & Qian, C., (2019). GNSS/INS/LiDAR-SLAM integrated navigation system based on graph optimization. *Remote Sens., 11*, 100.
21. Nahid, P., Michel, K., Hovannes, K., D'Amours, C., & Burak, K., (2022). A tutorial on AI-powered 3D deployment of drone base stations: State of the art, applications and challenges, *Vehicular Communications, 36*, 100474, ISSN 2214-2096. https://doi.org/10.1016/j.vehcom.2022.100474.
22. Annalisa, T. T., Thomas, A. B., & Todd, D. M., (2021). Active learning in robotics: A review of control principles. *Mechatronics, 77*, 102576, ISSN 0957-4158.
23. Aqel, M. O. A., Marhaban, M. H., Saripan, M. I., & Ismail, N. B., (2016). Review of visual odometry: Types, approaches, challenges, and applications. *SpringerPlus*.
24. Zhu, R., Yang, M., Liu, W., Song, R., Yan, B., & Xiao, Z., (2022). DeepAVO: Efficient pose refining with feature distilling for deep visual odometry. *Neurocomputing, 467*, 22–35.
25. An, D., Liu, J., Zhang, M., Chen, X., Chen, M., & Sun, H., (2020). Uncertainty modeling and runtime verification for autonomous vehicles driving control: A machine learning-based approach. *Journal of Systems and Software, 167*, 10.1016/j.jss.2020.110617.
26. Di Tomaso, E., & Baldwin, J. F., (2008). An approach to hybrid probabilistic models. *International Journal of Approximate Reasoning, 47*, 202–218.
27. Sitong, Z., Yibing, L., & Qianhui, D., (2022). Autonomous navigation of UAV in multi-obstacle environments based on a deep reinforcement learning approach. *Applied Soft Computing, 115*, 108194, ISSN 1568-4946, https://doi.org/10.1016/j.asoc.2021.108194.
28. Wortsman, M., et al., (2019). Learning to learn how to learn: Self-adaptive visual navigation using meta-learning. In: *2019 IEEE/CVF Conference on Computer Vision and Pattern Recognition (CVPR)* (pp. 6750–6759). IEEE, Piscataway, NJ.
29. Munir, S. A., & Jeon, M., (2021). SSTN: Self-supervised domain adaptation thermal object detection for autonomous driving. In: *2021 IEEE/RSJ International Conference on Intelligent Robots and Systems (IROS)* (pp. 206–213). Prague, Czech Republic. doi: 10.1109/IROS51168.2021.9636353.
30. Antonelo, E. A., Schrauwen, B., & Van, C. J., (2007). Generative modeling of autonomous robots and their environments using reservoir computing. *Neural Process Lett., 26*, 233–249.
31. Paz, L. M., Tardos, J. D., & Neira, J., (2008). Divide and conquer: EKF SLAM in O(n). *IEEE Trans. Robot., 24*, 1107–1120.
32. Holder, M., Hellwig, S., & Winner, H., (2019). Real-time pose graph SLAM based on radar. In: *Proceedings of the 2019 IEEE Intelligent Vehicles Symposium (IV)* (pp. 1145–1151). Paris, France.
33. Yang, Y., Tang, D., Wang, D., Song, W., Wang, J., & Fu, M., (2020). Multi-camera visual SLAM for off-road navigation. *Robot. Auton. Systs., 128*, 103505.

34. Yang, S., Scherer, S. A., Yi, X., & Zell, A., (2017). Multi-camera visual SLAM for autonomous navigation of micro aerial vehicles. *Robot. Auton. Syst.*, *93*, 116–134.
35. Memon, A. R., Wang, H., & Hussain, A., (2020). Loop closure detection using supervised and unsupervised deep neural networks for monocular SLAM systems. *Robot. Auton. Syst.*, *126*, 103470.
36. Lobos-Tsunekawa, K., et al., (2018). Visual navigation for biped humanoid robots using deep reinforcement learning. *IEEE Rob. Autom. Lett.*, *3*(4), 3247–3254.
37. Fayyad, J., Jaradat, M. A., Gruyer, D., & Najjaran, H., (2020). Deep learning sensor fusion for autonomous vehicle perception and localization: A review. *Sensors*, *20*, 4220. https://doi.org/10.3390/s20154220.
38. Goodfellow, Pouget-Abadie, J., Mirza, M., Xu, B., Warde-Farley, D., Ozair, S., Courville, A., & Bengio, Y., (2014). Generative adversarial nets. In: *Advances in Neural Information Processing Systems* (pp. 2672–2680).
39. Ignacio, C., Mariano De, P., Sen, W., Yvan, P., & Gerardo, G. A., (2018). Adaptive low-level control of autonomous underwater vehicles using deep reinforcement learning. *Robotics and Autonomous Systems, 107,* 71–86. ISSN 0921-8890. https://doi.org/10.1016/j.robot.2018.05.016.
40. Zhang, Y., Morel, O., Blanchon, M., Seulin, R., Rastgoo, M., & Sidibé, D., (2019). Exploration of deep learning-based multimodal fusion for semantic road scene segmentation. In: *Proceedings of the 14th International Joint Conference on Computer Vision, Imaging and Computer Graphics Theory and Applications* (pp. 336–343). SCITEPRESS—Science and Technology Publications: Prague, Czech Republic.
41. Jaedeug, C., & Kee-Eung, K., (2012). Nonparametric Bayesian inverse reinforcement learning for multiple reward functions. In: Bartlett, P., Pereira, F. C. N., Burges, C. J. C., Bottou, L., & Weinberger, K. Q., (eds.), *Advances in Neural Information Processing Systems* (Vol. 25, pp. 314–322).
42. Ho, J., & Ermon, S., (2016). Generative adversarial imitation learning. In*: Advances in Neural Information Processing Systems* (pp. 4565–4573).
43. Shopovska, I., Jovanov, L., & Philips, W., (2019). Deep visible and thermal image fusion for enhanced pedestrian visibility. *Sensors, 19*, 3727.
44. Han, S., & Xi, Z., (2020). Dynamic scene semantics SLAM based on semantic segmentation. *IEEE Access*, *8*, 43563–43570.
45. Gu, S., Lu, T., Zhang, Y., Alvarez, J. M., Yang, J., & Kong, H., (2018). 3-D LiDAR + monocular camera: An inverse-depth-induced fusion framework for urban road detection. *IEEE Trans. Intell. Veh.*, *3*, 351–360.
46. Shohei, C., & Hisayuki, S., (2021). Effectiveness of transfer learning in autonomous driving using model car. In: *2021 13th International Conference on Machine Learning and Computing (ICMLC 2021)* (pp. 595–601). Association for Computing Machinery, New York, NY, USA. https://doi.org/10.1145/3457682.3457773.
47. Jialin, S., & Yang, Q., (2010). A survey on transfer learning. *IEEE Transactions on Knowledge and Data Engineering, 22*(10), 1345–1359.
48. Aravind, B., Jaeyoung, L., Ashish, G., Krzysztof, C., & Sean, S., (2021). Transfer reinforcement learning for autonomous driving: From WiseMove to WiseSim. *ACM Trans. Model. Comput. Simul.*, *31*(3), 26. Article 15. https://doi.org/10.1145/3449356.

49. Aaron, F., Cynthia, R., & Francesca, D., (2018). *All Models are Wrong, but Many are Useful: Learning a Variable's Importance by Studying an Entire Class of Prediction Models Simultaneously.* arxiv:stat.ME/1801.01489.
50. Elbaz, G., Avraham, T., & Fischer, A., (2017). 3D point cloud registration for localization using a deep neural network auto-encoder. In: *IEEE Conference on Computer Vision and Pattern Recognition* (pp. 2472–2481). IEEE.
51. Engelcke, M., Rao, D., Wang, D. Z., Tong, C. H., & Posner, I., (2017). Vote3deep: Fast object detection in 3D point clouds using efficient convolutional neural networks. In: *IEEE International Conference on Robotics and Automation (ICRA)* (pp. 1355–1361). IEEE.
52. Caltagirone, L., Bellone, M., Svensson, L., & Wahde, M., (2019). LIDAR-camera fusion for road detection using fully convolutional neural networks. *Robot. Auton. Syst., 111*, 125–131.

Index

O

P

Q

R

Y

Z